Envisioning Networked Urban Mobilities

Envisioning Networked Urban Mobilities brings together scientific reflections on the relations of art and urban mobilities and artistic research on the topic. The editors open the book by setting out the concept grounded in the exhibition curated by Aslak Aamot Kjærulff and refers to earlier work on mobilities and art generated by the Cosmobilities Network. This third volume has two sections, both consisting of short papers and illustrations. The first section is based on artists who were part of the conference's art exhibition, and the second part is based on theoretical reflections on art and artists.

Aslak Aamot Kjærulff holds a PhD in Mobilities and Action Research from Roskilde University, Denmark. He currently organizes a transdisciplinary research and arts organization called Diakron and teaches at Roskilde University.

Sven Kesselring is Research Professor in Automotive Management: Sustainable Mobilities and the director of the Master of Science program Sustainable Mobilities at Nürtingen-Geislingen University (NGU), Germany. His research focuses on the sociology of (auto)mobilities, social theory, and the impact of technology and digitalization on everyday and professional lives. He is the founder and co-manager of the international Cosmobilities Network and co-director of the joint PhD program Sustainable Mobility and Mobility Cultures of TU Munich and NGU. He is co-founder and co-editor of the new journal *Applied Mobilities* and Studies in Mobility and Transport. He has edited several books including *Aeromobilities* (with Saulo Cwerner and John Urry).

Peter Peters is Assistant Professor in the Department of Philosophy in the Faculty of Arts and Social Sciences at Maastricht University, the Netherlands. He is trained as a sociologist and holds a PhD for his dissertation on mobilities in technological cultures, in which he combines insights from social theory and science and technology studies to analyze practices of travel.

Kevin Hannam is Professor of Tourism Mobilities in the Business School at Edinburgh Napier University, UK, and a research affiliate at the University of Johannesburg, South Africa. In 2015/2016 he was awarded a Vice-Chancellor's International Scholarship to the University of Wollongong, Australia. Previously he was at Leeds Beckett University, UK, and the University of Sunderland, UK.

Networked Urban Mobilities Series

Editors: Sven Kesselring
Nürtingen-Geislingen University
Malene Freudendal-Pedersen
Roskilde University

The Networked Urban Mobilities series resulted from the Cosmobilities Network of mobility research and the Taylor & Francis journal *Applied Mobilities*. This three-volume set, ideal for mobilities researchers and practitioners, explores a broad number of topics including planning, architecture, geography, and urban design.

Exploring Networked Urban Mobilities
Theories, Concepts, Ideas
Edited by Malene Freudendal-Pedersen and Sven Kesselring

Experiencing Networked Urban Mobilities
Practices, Flows, Methods
Edited by Malene Freudendal-Pedersen, Katrine Hartmann-Petersen and Emmy Laura Perez Fjalland

Envisioning Networked Urban Mobilities
Art, Performances, Impacts
Edited by Aslak Aamot Kjærulff, Sven Kesselring, Peter Peters and Kevin Hannam

"In the emerging field of artistic research, sensory and embodied ways of knowing take center stage. At the same time, qualitative social scientists and geographers are developing a keen interest in new methodologies that draw inspiration from the arts. This timely volume on the nexus of arts and mobilities research stages an engaged dialogue between leading academic scholars and artists."

Henk Borgdorff, Academy of Creative and Performing Arts, Leiden University, the Netherlands

Envisioning Networked Urban Mobilities

Art, Performances, Impacts

Edited by
Aslak Aamot Kjærulff, Sven Kesselring, Peter Peters and Kevin Hannam

Routledge
Taylor & Francis Group
LONDON AND NEW YORK

First published 2018 by Routledge

2 Park Square, Milton Park, Abingdon, Oxfordshire OX14 4RN
52 Vanderbilt Avenue, New York, NY 10017

Routledge is an imprint of the Taylor & Francis Group, an informa business

First issued in paperback 2019

Library of Congress Cataloging-in-Publication Data
A catalog record for this book has been requested

ISBN: 978-1-138-71236-2 (hbk)
ISBN: 978-0-367-33180-1 (pbk)

Typeset in Sabon
by Apex CoVantage, LLC

To John

Contents

Contributors

Ulrike Boskamp is a German art historian. She studied art history and philosophy at Freie Universität Berlin, Germany, has worked at Deutsches Forum für Kunstgeschichte in Paris in 1999/2000, and was an assistant professor in art history at Freie Universität Berlin 2006 to 2010. Her PhD on color in art, art literature, and science in eighteenth-century France, *Primärfarben und Farbharmonie*, was published in 2009. Since 2011, she has been engaged in the research project *On the Border: Travelling Artists Accused of Espionage (17th–19th Centuries)* as a senior fellow in the research group Transcultural Negotiations in the Ambits of Art at Freie Universität Berlin, funded by the German Research Foundation (DFG). Departing from recurring accounts about the confusion of artists with spies in art literature, artist's biographies, and anecdotes, it tackles questions about the military power ascribed to images of landscapes and to (male) traveling artists in Europe, the status of images in military espionage and surveillance, steganographic coding of artistic images, the status of visualizations of and in contested border regions, and the role and status of this topical narration.

Mike Collier is a lecturer, writer, curator, and artist. He studied fine art at Goldsmiths College, London, UK, and is currently Reader in Fine Art at the University of Sunderland, UK. Much of his work is based around walking—through the city, the countryside, and urban Edgelands. His work pays close attention to the environment through which he walks and is usually place-specific. He integrates image and text, often drawing on the poetic qualities of colloquial names for places, plants, and birds. He has shown in the UK and abroad and his work is in a number of public and private collections. In 2010 he co-founded WALK (Walking, Art, Landskip and Knowledge), a research center at the University of Sunderland which looks at the way we creatively engage with the world as we walk through it. As an artist and curator, he has been responsible for a number of high-profile exhibitions under the auspices of WALK including co-curating *Walk On: From Richard Long to Janet Cardiff—Forty Years of Art Walking*, an exhibition which toured the UK in 2013/2014; and *Wordsworth and Bashō: Walking Poets*, an exhibition of manuscripts by William and Dorothy Wordsworth and Matsuo Bashō shown alongside newly commissioned work by 22 leading contemporary artists from the UK and Japan, held at Dove Cottage in 2014 and Kyoto, Japan, in 2016.

Angie Cotte is a French/American binational and has been actively contributing to cultural networks, advocacy platforms, and cooperation projects in Europe since the early 1990s. She worked for the Eurocreation Agency, the European Pépinières residency program, and participated in Policymakers meetings in European Capitals of Culture. Her role has been to start up projects and develop them into fully fledged cultural networks by increasing membership, developing funding partnerships, and organizing conferences and seminars. She currently manages the activities of the Roberto Cimetta Fund and raises awareness within institutions on artistic and cultural mobility practices and needs in Europe, the Arab world, and the Middle East. She develops studies on mobility trends and advocates for more recognition of the role of arts and culture in the external relations of the European Union, particularly for the neighborhood region. She holds a master's degree from Sorbonne University, Paris, France, in Intercultural Relations and a bachelor's degree from Goldsmiths College, London, UK, in Theatrical Studies. She lectures at the Universities of Bobigny and Senart on funding opportunities for grassroots cultural projects that address intercultural challenges. She is currently a member of the Executive Committee of Culture Action Europe, the Brussels-based advocacy platform for culture and the arts in Europe.

Kevin Hannam is Professor of Tourism Mobilities in the Business School at Edinburgh Napier University, UK, and a research affiliate at the University of Johannesburg, South Africa. In 2015/2016 he was awarded a Vice-Chancellor's International Scholarship to the University of Wollongong, Australia. Previously he was at Leeds Beckett University, UK, and the University of Sunderland, UK, as Associate Dean (Research). He is a founding co-editor of the journals *Mobilities* and *Applied Mobilities* (Routledge), co-author of the books *Understanding Tourism* (Sage) and *Tourism and India* (Routledge), and co-editor of the *Routledge Handbook of Mobilities Research*, *Moral Encounters in Tourism* (Ashgate), *Tourism and Leisure Mobilities*, and *Event Mobilities* (both with Routledge). He has extensive research experience in South and Southeast Asia and has published in many leading journals. He has a PhD in Geography from the University of Portsmouth, UK, and is a Fellow of the Royal Geographical Society (FRGS), member of the Royal Anthropological Institute (RAI), and Vice-Chair of the Association for Tourism and Leisure Education and Research (ATLAS).

Antonia Hernández is a media artist and SSHRC-supported PhD student in Communication Studies at Concordia University. Mixing media practice and theoretical research, her interests involve the domestic side of digital networks, post-digital labor, and migration. She is a co-founding member of MAT3RIAL, a collective of researchers, designers, and developers.

Michael Hieslmair studied Architecture at Graz Technical University, Austria, and Delft Technical University, the Netherlands. Since 2005 he has collaborated with Michael Zinganel as curators, writers, and artists on projects about urban and transnational mobility and migration. In 2012 they co-funded the research platform Tracing Spaces. From 2014 to 2016 both had been research fellows at the Academy of Fine Arts Vienna and heads of research of the project *Stop & Go: Nodes of Transformation and Transition*. In 2015 they published (with Tarmo Pikner) 'Test Run—Stop and Go. Mapping Nodes of Mobility and Migration,' in Judith Laister and Anna Lipphardt (eds.), *Urban Place-Making Between Art, Qualitative Research and Politics. Anthropological Journal of European Cultures*, Volume 24, No. 2, 2015, 117–127. In 2016 they co-edited a special issue of the magazine *dérive—Zeitschrift für Stadtforschung*, No. 63, 2016. Learn more about their work at www.mhmz.at, www.tracingspaces.net, and www.stopandgo-transition.net.

Sven Kesselring is Research Professor in Automotive Management: Sustainable Mobilities at Nürtingen-Geislingen University, Germany. His research focuses on the sociology of (auto)mobilities, social theory, and the impact of technology and digitalization on everyday and professional lives. He is the founder and co-manager of the international Cosmobilities Network. From 2014–2016 he was vice-president of the International Association for the History of Transport, Traffic and Mobility (T2M). He is co-founder and co-editor of the new journal *Applied Mobilities* (Taylor & Francis). He has edited several books including *New Mobilities Regimes: The Analytical Power of the Social Sciences and the Arts* and *Aeromobilities*.

Aslak Aamot Kjærulff holds a PhD in Mobilities and Action Research from Roskilde University, Denmark. He currently organizes a transdisciplinary research and arts organization called Diakron and teaches at Roskilde University. The core trajectory of his practice is to trace how concepts and ideas travel across cultures and disciplines. Most of the research he takes part in is carried out in ways that engage several disciplinary and organizational backgrounds at once. This involves exploratory methodological designs, where problem formulations, research processes, and artistic productions are invented collaboratively. The outcomes of this relational practice have been exhibitions, government reports, artworks, transdisciplinary communities of practice, radio documentaries, and fictional as well as academic texts. Ultimately this means engaging with and contributing to several types of organizations and public spaces.

Annette Kranen is a researcher in Early Modern European art history and a member of the DFG-funded research project *Transcultural Negotiations in the Ambits of Art* based at Freie Universität Berlin, Germany. She has worked on the visual cultures of travel and early anthropology.

Her master's thesis on the images made during James Cook's third journey to the Pacific was completed in 2013. In 2013/2014 she was Assistant Professor at the Department of Art History at Freie Universität Berlin. Currently she is preparing her PhD thesis on drawings and etchings made by European travelers to the Ottoman territories in the seventeenth and early eighteenth centuries. Her research in this field is focusing on the tensions between the travelers' knowledge about the ancient and biblical sites on the one hand and the actual state of the places by the time they arrived there on the other. This approach aims at considering transcultural encounters in a diachronic sense, thus reconstructing a historical dimension of the 'contact zone.'

Lee Lee is a visual artist who cultivates 'creative ecologies' inspired by natural ecologies to explore human intersections with the natural world. Starting with figurative and landscape painting, her work grows into community engagements that address social and ecological issues presented by a post-industrial chemical age. Her practice settles in remote places in order to understand the broader impacts of the way we live now. The methodology that has evolved through carrying the Debris Project across the world is now informing an emerging collaborative practice around seeds. Intimate relationships with the landscape has been paramount to Lee Lee's work and she has dedicated her career to finding ways her creative acts may support wildlands. After curating the *SEED :: Disperse* exhibit at the Dairy Arts Center in Boulder, Colorado, USA, she is cultivating collaborations that focus on food security and wildlands restoration in the Rocky Mountains of Colorado and New Mexico, the North Woods of Maine, and in the Red Zones of Port-au-Prince, Haiti. Recently this was realized as a creative exchange with Irish artists, initiated by Rian Kerrane, where artists were asked to explore the connections between Ireland and the West through a pair of residencies and exhibitions at RedLine in Denver and the Regional Cultural Center in Letterkenny. At a time when Ireland is shifting towards industrial meat production, this was an opportunity to weave her history on the Flanigan's Cattle Ranch in Colorado with a long involvement in the Slow Food Movement to look at both the industrial impacts and potential opportunities of sustainable meat production.

Bart Magnus studied Germanic Languages (Dutch and English literature) and teacher training and has a master's degree in Theatre Studies. In 2009, after two seasons at Het Theaterfestival (the first of which as a trainee), he started working at VTi (Vlaams Theater Instituut—Flemish Institute for the Performing Arts) as a researcher on the European SPACE project (www.arts-mobility.info). SPACE stands for Supporting Performing Arts Circulation in Europe. He worked mainly on the subproject Travelogue, bringing together and researching dispersed and fragmented data collections about international touring in the performing arts. In July 2010 he was appointed head of performing arts documentation at VTi, which merged into Flanders Arts Institute (www.flandersartsinstitute.be) in January 2015. During this period he was responsible for the (analog and digital) documentary collection as well as for the database containing both collection metadata and data on Flemish performing arts productions. Since August 2016 he works at PACKED, center of expertise in digital heritage. He is also a board member of the European Network of Information Centres on the Performing Arts (ENICPA—www.enicpa.info). These experiences allowed him to build up expertise on Flemish and international (performing) arts and more specifically on the intersection between databases, research, and documentary/archival collections.

María Paz Peirano is Lecturer in Film Studies at Universidad de Chile, with a PhD in Social Anthropology from the University of Kent, UK. She trained as a social anthropologist at Universidad de Chile and holds postgraduate degrees in both Documentary Film and Film Studies from Universidad de Chile and Pontificia Universidad Católica de Chile, respectively. Her research involves an ethnographic approach to film as social practice, focusing on film festivals and the construction of contemporary Chilean cinema in transnational settings. She has recently worked as a Postdoctoral Researcher at Leiden University, the Netherlands, studying the impact of international festivals on Latin American cinema, and is co-editor of the volume *Film Festivals and Anthropology* (Cambridge Scholars, 2017), which explores the relationship between ethnography, film festivals, and visual anthropology.

Peter Peters is Assistant Professor in the Department of Philosophy in the Faculty of Arts and Social Sciences at Maastricht University, the Netherlands. He is trained as a sociologist and holds a PhD for his dissertation on mobilities in technological cultures, in which he combines insights from social theory and science and technology studies to analyze practices of travel. From 2008 to 2013 he was a professor in the research center 'Autonomy and the Public Sphere in the Arts' at Zuyd University, Maastricht, the Netherlands. Here, he developed research on artistic research and its relation to science and technology studies, as well as site-specific art, and art in the public sphere. His current research focuses on how knowledge, techniques, and craftsmanship are developed in historical and contemporary practices of pipe organ building. Peters is author of *Time, Innovation and Mobilities* (Routledge, 2006). He has published his work in journals including *Sound Studies*, *Interdisciplinary Science Reviews*, *Research Policy*, *Configurations*, and *Mobilities*. He is a member of the editorial board of *Mobilities* (Sage) and review editor of the journal *Time & Society* (Sage).

David Pinder is Professor of Urban Studies at Roskilde University, Denmark. His research explores how urban spaces are produced, imagined, performed, and contested. He has written widely on utopianism and cities, and on efforts to re-imagine and reconstitute the possibilities of urban society within and against capitalist urbanization. That has been with specific reference to the ideas and spatial practices of modernist and avant-garde movements, including the situationists, as well as strands of critical urban theory. He also focuses on contemporary artistic explorations, interventions, and urban politics, including the relation to psychogeography, walking art, and radical cartographies. Among his publications is the book *Visions of the City: Utopianism, Power and Politics in Twentieth-Century Urbanism* (2005), and he has guest-edited the special issue of *Cultural Geographies* on *Arts of Urban Exploration*. Before moving to Roskilde, he taught at Queen Mary University of London, UK, and he has also held visiting positions at Princeton University and at City University of New York Graduate Center, USA.

Nikki Pugh is a British artist whose practice focuses on interactions between people and place. Exploring questions relating to how we perceive, move through, and interact with our surroundings, she harnesses various tools and techniques adopted from walking-based practices, guided tours, physical computing, locative media, pervasive gaming, installation, and collaboration. Her main areas of research include making, prototyping and participatory playtesting as tools for—and sites of—knowledge production. At the time of writing, she is a (returning) fellow at Birmingham Open Media, and also an associate artist with Fermynwoods Contemporary Art. Recent residencies include those based at Wolverhampton School of Art, Pervasive Media Studio (Bristol, UK), Coventry Transport Museum, and Birmingham Museum & Art Gallery. She holds undergraduate degrees in Materials Science (Birmingham University, UK) and Art & Design (University of Central England, UK).

The ongoing project *Colony* was featured in *Right Here, Right Now*—The Lowry's "major exhibition providing a thought provoking snapshot of contemporary digital art." Subsequent projects include *Orrery for Landscape, Sinew and Serendipity*, which investigates questions about the physical and emotional experiences of cycling (and of being the person left at home); the frictions of data visualization; and what happens when you shift from thinking about markers on a map to an awareness of the changing rhythms of effort and terrain.

Jen Southern is an artist and Lecturer in Fine Art at Lancaster University, UK. As director of the mobilities lab at the Centre for Mobilities Research she is involved in developing and supporting experimental mobile and creative research methods. Her transdisciplinary research explores collaborative uses of GPS technology, both producing and making visible a sense of 'comobility,' of being mobile with others at a distance. With over 20 years' experience as an internationally exhibiting artist, her recent publications include *Leonardo Electronic Almanac* and *The Canadian*

Journal of Communications, with recent commissions from In Certain Places, Abandon Normal Devices Festival, National Football Museum, UK, and the Mobile Media Studio, Montreal, Canada. Her work can be seen at theportable.tv.

Samuel Thulin is a SSHRC postdoctoral fellow at the Centre for Mobilities Research at Lancaster University, UK, and a sound artist currently based in Montreal, Canada. He holds a PhD in Communication Studies from Concordia University. Combining theory and creative practice, Thulin investigates intersections in sound studies, media studies, and mobilities research. He has conducted research in the areas of mobile music, sound mapping, and locative audio, as well as writing on methodologies that blend practice and theory, such as research-creation and the concept of resonance. His postdoctoral research explores the ongoing proliferation of mobile audio tools and practices, and develops the concept of 'situated composition' as a way of understanding composition as collaborative, distributed, and emergent, entailing not only sound, but social, material, virtual, and digital combinations. He is currently editing a special issue of *Wi: Journal of Mobile Media* focused on *Mobile Making*. Thulin's artworks exist variously as punctual performances and installations in physical spaces, as online creations, and as geolocated compositions that rely on movement through places in order to be activated. In each case, the contingency of place features as a vital part of the creative process. Thulin has shared his academic and artistic work via scholarly publications, and at venues in Europe and the Americas.

Michael Zinganel studied Architecture at Graz Technical University, Austria, Art at the Jan van Eyck Academy Maastricht, the Netherlands, and History at the University of Vienna, Austria. He is currently teaching at Bauhaus Dessau Foundation and Vienna University of Technology. Since 2005 he has collaborated with Michael Hieslmair as curators, writers, and artists on projects about urban and transnational mobility and migration. In 2012 they co-funded the research platform Tracing Spaces. He produced and co-edited *Holiday after the Fall—Seaside Architecture and Urbanism in Bulgaria and Croatia* (with Elke Beyer and Anke Hagemann, 2013). From 2014 to 2016 Hieslmair and Zinganel had been research fellows at the Academy of Fine Arts Vienna and heads of research of the project *Stop & Go: Nodes of Transformation and Transition*. In 2015 they published (with Tarmo Pikner) 'Test Run—Stop and Go. Mapping Nodes of Mobility and Migration,' in Judith Laister and Anna Lipphardt (eds.), *Urban Place-Making Between Art, Qualitative Research and Politics. Anthropological Journal of European Cultures*, Volume 24, No. 2, 2015, 117–127. In 2016 they co-edited a special issue of the magazine *dérive—Zeitschrift für Stadtforschung*, No. 63, 2016. Learn more about their work at www.mhmz.at, www.tracingspaces.net, and www.stopandgo-transition.net.

Preface

'Networked Urban Mobilities': This was the title for the conference and art exhibition that the Cosmobilities Network organized from the 5th to the 7th of November 2014 in Copenhagen. It was a big scientific event and the reason for it was the tenth anniversary of the international research network. The conference was jointly organized by a team from Aalborg University and Roskilde University.

About 160 participants came to Aalborg University's campus in Copenhagen and in this three-volume set many of them are presenting their work and ideas. The book you hold in your hand is the third volume and it collects contributions from the conference on 'Art, Performances, Impacts' as the subtitle tells. By so doing it propels a topic which has been a significant element of the work within the Cosmobilities Network since the 2008 conference on mobility and art in Munich: to demonstrate the analytical power of art and social science. Volume one is edited by Malene Freudendal-Pedersen and Sven Kesselring and compiles theoretical debates, conceptual considerations, and new ideas and perspectives which are prominent within the new mobilities paradigm. Volume two is edited by Malene Freudendal-Pedersen, Katrine Hartmann-Petersen, and Emmy Laura Perez Fjalland and presents a wide range of contributions which have the ambition to illustrate the strength of and variety in subjects and empirical broadness within the mobilities paradigm. Its subtitle indicates this in a pithy way: 'Practices, Flows, Methods.'

The Conference in Copenhagen was a special event and a milestone for the network in many ways.

Firstly, the Cosmobilities Network celebrated its birthday. Ten years before a small crowd of people came together in Munich for a workshop on 'Mobility and the Cosmopolitan Perspective.' At this time no one considered this the birth of a long-lasting collaboration and research network. The name Cosmobilities was mostly a running gag on how to shorten cosmopolitan mobilities during these days. At some point John Urry laughingly said, 'This name calls for a network!' And more than a decade later, the Cosmobilities Network plays a substantial role within the mobilities turn in social science and beyond. At the tenth anniversary conference we were celebrating

Cosmobilities as an academic space, a place for encounters, and a synonym for cutting-edge research and scientific innovation. A huge number of individual scholars and research institutions worldwide have generated a new interdisciplinary literature and a new thinking on the social transformations of the modern mobile world, its risks and opportunities. The 'new mobilities paradigm' has influenced work and thoughts of academic scholars as well as practitioners in public authorities, industry, and civil society.

Secondly, what we luckily didn't know at this time, it was the last Cosmobilities conference where two very important academic personalities and thinkers, who both played an important role in the 2004 workshop and the beginning of the network, were still with us.

German sociologist Ulrich Beck was invited to the 2004 workshop and in the aftermath he fostered the founding process of the Cosmobilities Network. In the 2014 Conference catalogue he wrote a welcome note to the participants, stating that Cosmobilities "has become a reflexive place and space for re-thinking the basic principles of modernity and for the future of modern societies."

John Urry's role since 2004 and up until a very sad day in March 2016 cannot be overestimated. In many ways he was and still is the 'spiritus rector' and the mentor of the network and of many, many mobilities scholars. Without his unique personality and his brilliant mind the network would not be what it is today. And this is said without any exaggeration. For the 2014 conference catalogue he wrote:

> Throughout the last decade Cosmobilities has provided a really brilliant space that has nurtured the emerging mobilities paradigm. As a horizontal network of many senior and junior colleagues you have done a great job in bringing together scholars from many different fields and theoretical approaches as well as research traditions. You have been bridging the gap between academia and practitioners, too. And hopefully Cosmobilities will long continue.

The book series *Networked Urban Mobilities* is dedicated to these two thinkers. Both of their words we consider as the assignment and the mission of the Cosmobilities Network. We hope that the book in your hand and all three volumes together will give an overview of the depth, the diversity, and analytical sharpness of the new mobilities paradigm and the potentials of the scholars of the network. Beyond this we wish you an exciting and illuminating reading experience.

We would be glad if this book aroused some interest in our work, and maybe we will see you at one of the next Cosmobilities conferences.

Malene Freudendal-Pedersen, Sven Kesselring
Copenhagen, September 2016

1 Networked Urban Mobilities

Art, Performances, Impacts

Kevin Hannam, Aslak Aamot Kjærulff, Peter Peters, and Sven Kesselring

In his final book *What Is the Future?*, John Urry (2016) discussed possible urban futures and outlined four possible scenarios based upon his reading of the mobilities literature. Firstly, he notes the possibility of the 'fast mobility city' where people experience more intensive mobile lives, giving the examples of Dubai, Hong Kong, and Singapore, among others, as examples of this blueprint. Such cities would be networked for the elite and "hum with talk" as people meet in 'coffices' for choreographed meetings. Driverless cars, helicopters, and drones would enable the elite to make seamless connections. But, he notes, this type of city can only be made if a new post-carbon energy system is innovated and implemented. Secondly, he develops the scenario of the 'digital or smart city' where the physical movement of people and things becomes replaced by digital communications and experiences and travel as such slows down. The 2015 European Digital City index lists London, Amsterdam, and Stockholm as the leading three 'digital cities' in Europe (EDCi, 2015). Digital lives would mean that digital experiences and encounters would be as good as the 'real thing' as environments become smarter and technologies such as 3D printing become more ubiquitous. He notes that for younger people in the Global North smartphones have already begun to replace cars as status symbols and cars themselves have integrated with smartphones to become 'smartcars.' As a result, cities may become more dispersed but dependent on virtual networks. With the advent of the 'digital city' though comes new vulnerabilities to system breakdowns.

Thirdly, Urry discusses the possibility of the 'liveable city.' The top three 'liveable' cities according to *The Economist* (2016) are Melbourne, Vienna, and Vancouver based upon safety, health care, educational resources, infrastructure, and the environment and tend to be mid-sized cities in wealthier countries (Damascus is bottom is this index). The liveable city, Urry notes, is based upon environmental innovations rather than technological developments that are a feature of fast or digital cities but may include 'green' developments such as solar energy (see Beatley, 2007). Cities would become fragmented into self-sufficient neighborhoods with people moving by bicycle and other 'light' modes of transport such as collectively owned electric vehicles. Fourthly, Urry finally discusses the possibility of the 'fortress city' where gated societies are taken to the extreme and the elite live in 'defensible' enclaves (Killgren, 2015). Those outside these enclaves would inhabit the 'wild' and there would be increased conflict over scarce resources in a *Mad Max*- type dystopia. Cities with centers of power such as Washington, DC, Paris, or Ankara may become the fortresses at the frontline for cyber and terrorist attacks such that individual civil liberties become increasingly circumscribed as the population is stopped and searched for the 'common good.'

Overall, Urry (2016) helps us to distill what we may want our common urban future to look like and this book seeks to further these debates by investigating how artistic practices can contribute to shaping new imaginaries of networked urban mobilities, as well as researching these mobilities in novel ways. Historically, the city has always been an important site for artistic production and reflection. From the impressionist paintings in nineteenth-century Paris to artistic interventions in public spaces in New York City in the 1960s to twenty-first-century community art projects in Rio de Janeiro, artists have worked with urban experiences and sought new ways to engage with the problems and politics of city spaces. Recently, some of this work resonates with the emerging field of artistic research in which subjective, sensory, and embodied ways of knowing take center stage. Artistic research practices result not only in art objects and performances being presented to art worlds, but also in epistemic claims that are contested in academic communities (Borgdorff, 2012). As such it takes place at universities and in laboratories, and it unfolds at exhibitions and during performances.

Mirroring this development in the arts is the interest of social scientists in new arts-based methods as a way to go beyond the boundaries of academic modes of knowing (Law & Ruppert, 2016). Through mutual involvement between social sciences and art practices, these methods draw on artistic repertoires in organizing autonomy, questioning routines and breaching social conventions. These repertoires deal with ambiguities, materiality, performativity, site-specificity, and networked events in different ways than conventional academic research practices.

Envisioning Networked Urban Mobilities: Art, Performances, Impacts thus seeks to engage with various artistic practices and social science methodological debates that 'envision' the urban future because relatively little critical work has been published on what happens when artistic works and artists become mobile. This book seeks to go beyond the dualism between art and science, engaging with science and technology studies to consider 'matters of concern' for new strategies to re-envision cities.

Following this introduction, the book opens (Chapter 2) with a philosophical discussion by Aslak Aamot Kjærulff about the need to develop dialogues that re-negotiate the divisions between art, social science, and science and technology—a central theme of the book as a whole. He notes the importance of 'lower theories' rather than 'grand narratives' as a way of engaging with everyday understandings of how life might be lived and researched. Curating conversations, he offers, is a particular methodological practice that might enable us to ask difficult and produce new questions about the future shape of our urban networks. Such conversations may leave something behind that stimulates change and leaps of imagination, thus cunningly outwitting what has been taken for granted as normal.

Jen Southern in Chapter 3 takes this theme up by developing the notion of 'reflective assemblages' and 'comobility' through locative media art. She discusses the mobile phone app *Comob Net* as a way of researching the use of collaborative GPS technologies in everyday mobilities. This involves participants becoming 'knowingly entangled' in problematic situations that seek to develop and stretch their geographical imaginations as they become engaged with coproducing her art. *Comob Net* is thus simultaneously art practice, speculative design, and a mobile sociological method involving the co-creation of (sometimes uncomfortable) shared knowledge: "Working with methods that are simultaneously *material, live, and relational* allows participants to not only be 'informants' but also to be analysts of their own activities and data through collaborative annotation, and in artworks through live and participatory processes that draw people together for discussion."

Chapters 4, 5, and 6 by Peter Peters, Samuel Thulin, and David Pinder, respectively, further develop these philosophical and methodological insights through an emphasis on methods and practices of listening to sonic mobilities, rhythms, and acoustic ecologies (see also Revill, 2013). As Thulin notes: "We are always hearing mobility, though we may not always be actively listening to it." Peter Peters in Chapter 4 examines sonic and theatrical artistic interventions as a contribution to the development of mobile methods through a discussion of the work of the Italian artist Daniela de Paulis and the German theater group Rimini Protokoll. He argues that these artists allow us to investigate the "systemic gestures of transportation and mobilities that stretch out from urban fabrics to connect them over large distances" such that we may come to know a system from the inside. Like Southern, Peters also acknowledges the importance of the audience in the co-creation of new understandings of the use of urban spaces.

In Chapter 5 Samuel Thulin investigates the playful artistic intervention of a music-route in Montreal, Canada. Again, he focuses on practices of co-creation through location-based media as an opportunity for participants to modulate their habitual movements through urban space. He concludes that "sonic and mobile practices are increasingly networked and interdependent, pointing to the importance of sound and listening for mobilities research." Meanwhile, in Chapter 6 David Pinder reflects on the 'spectral' sounds of motorways by engaging with artists that have sought to listen to roads beyond their 'unreadable surfaces.' Roads may be reanimated physically by flora and fauna and more implicitly by dispersed memories (Crang, 1996) or 'ghosts' of past political struggles against road building. In particular he discusses the London walking art project *Linked* by Graeme Miller, which seeks to reclaim this ecology through encounters that are always sonic as well as tactile and visual experiences.

Developing the theme of walking as both artistic practice and methodology, Mike Collier and Kevin Hannam in Chapter 8 reflect on artistic responses to the Walking with Wordsworth and Bashō exhibition held at Dove Cottage in the Lake District, UK. They show how these poets are linked through the agency of walking with a wider context of anti-urbanism and then discuss contemporary artistic responses to these poets. They seek to develop the methodology of 'meanders' as a way of developing the co-creation of artistic experiences through walking in both urban and non-urban networks. In Chapter 11, Ulrike Boskamp and Annette Kranen also develop an analytical view on artistic mobility and history through their examination of the mobility entanglements of representations of the Dardanelles straits originating from two late seventeenth-century French journeys to the Ottoman Empire. They emphasize the complex systems of mobility and technology that allowed representations of the Dardanelles landscapes and ecologies to be constructed. A visual investigation of infrastructural hubs and landscapes have been embedded in a vehicle by Michael Hieslmair and Michael Zinganel. In Chapter 10 they provide written and visual access to their ways of working with one of Europe's major transport corridors, from Estonia through Austria to Bulgaria and Turkey. By combining visual mapping, sculptural public interventions, and artistic workshop formats, they point to functional and aesthetic qualities of infrastructural nodes and the lived experiences that pass through them.

Ecologies, technologies, and the metaphor of colony are reflected upon in Chapter 7 by the artist Nikki Pugh in her attempt to open up a conversation about her co-created works. Using the long-running and multifaceted research project *Colony* as a series of jumping-off points, she outlines her underlying methodological approach towards empirical knowledge production through her inquiry-led practice. Her technological sculptural devices assemble physical computing, body-awareness, and performance to act as experience-based prompts for insight, conversations, and reflection on contemporary urban ecologies. As she puts it, her aim is to "make a family of landscape-aware 'creatures' that are carried across the city." These 'creatures' interact with those that carry them as she develops the methodology of 'playtesting' participation. Such experimentation is also developed by Antonia Hernández in her reflections on her 'performative fungal strategies' in Chapter 9. By examining how mold performs under a microscope she develops a subtle artistic critique of living within digital networks and the ways in which human ecologies will interact with such digital spaces in the future, emphasizing the entanglements of human and non-human mobilities. Lee Lee develops this critical analysis of the relations between the human and non-human in Chapter 12 through a discussion of the Debris Project—a traveling artistic response to the circulation of plastic pollution debris in the seas. She draws directly on her conversations with John Urry about this project and its significance for the future in terms of inspiring future environmental education through the co-creation of artistic projects with children.

The mobilities of traveling artists is the final theme that is central to this book and María Paz Peirano discusses this in depth through her analysis of film mobilities and the construction of Chilean cinema in Chapter 13. Film circulation and the international film festival network, she argues, has been important in the reconfiguration of alternative urban networks and their representations. As Tzanelli (2016) has argued, cinema has been significant in the re-imagining of contemporary urban spaces and will continue to be in the future. Chapters 14 and 15 by Angie Cotte and Bart Magnus, respectively, develop arguments for further artistic and cultural mobilities so that that we do not end up in a world of fortress cities unable to engage with the global insecurities and inequalities which the methodologies developed in this book seek to confront.

References

Beatley, T. (2007). Envisioning Solar Cities: Urban Futures Powered by Sustainable Energy. *Journal of Urban Technology*, 14(2), 31–46.

Borgdorff, H. (2012). *The Conflict of the Faculties: Perspectives on Artistic Research in Academia*. Leiden: Leiden University Press.

Crang, M. (1996). Envisioning Urban Histories: Bristol as Palimpsest, Postcards, and Snapshots. *Environment and Planning A*, 28(3), 429–452.

The Economist. (2016). The World's Most Liveable Cities. Available at: www.economist.com/blogs/graphicdetail/2016/08/daily-chart-14 Accessed 21 August 2016.

EDCi. (2015). *European Digital City Index*. Available at: https://digitalcityindex.eu/ Accessed 21 August 2016.

Killgren, L. (2015). Safety Obsession Creates Risk of 'Fortress' Cities. *Financial Times* (online). Available at: www.ft.com/cms/s/2/680d8d2e-d897-11e4-ba53-00144feab7de.html#axzz4Hu5HwSAf Accessed 21 August 2016.

Law, J., and Ruppert, E. (Eds.) (2016). *Modes of Knowing: Resources from the Baroque*. Manchester: Mattering Press.

Revill, G. (2013). Points of Departure: Listening to Rhythm in the Sonoric Spaces of the Railway Station. *Sociological Review Monograph*, 61, 51–68.

Tzanelli, R. (2016). *Thanatourism and Cinematic Representations of Risk: Screening the End of Tourism*. London: Routledge.

Urry, J. (2016). *What Is the Future?* Cambridge: Polity.

2 Curating Conversations

Reformulating Questions in Mobilities Arts and Research

Aslak Aamot Kjærulff

This chapter is based on conversations with contributors to the 'Mobile Exhibition' at the ten-year anniversary conference of the Cosmobilities Network. The text is an attempt to further the dialogues between artistic and academic practitioners who work with mobilities. Four main threads are spun in this article. Firstly, I argue that the future of mobilities research depends on the ability to create ideas that can permeate or re-negotiate the current divisions of practices into disciplines and sectors. Secondly, I seek to justify that the ambitions of 'lower theories' to operate beneath or between various different disciplines might be worth examining closer, in a time where grand narratives, such as the Anthropocene era or global urbanization, are occupying common horizons. Thirdly, I emphasize that curating and hosting dialogues could be a format to be taken much more seriously in the contexts of the former two endeavors. And finally, the text ends with a short manifesto of three questions and statements that aim to provoke an interest in processual question formulation.

> There are four stages of accepting scientific theory:
>
> i) this is worthless nonsense;
> ii) this is an interesting, but perverse point of view;
> iii) this is true, but quite unimportant;
> iv) I have always said so.
>
> (J.B.S. Haldane, in Sally O'Reilly, Introduction to Macuga, 2008: 9)

During the summer of 2010 a conversation was staged at Centre Pompidou in Paris. The topic was the question 'who owns space and time?' The event was a re-enactment of a conversation between the philosopher Henri Bergson and the physicist Albert Einstein. Here, Einstein infamously dismissed Bergson's notions of time and space as merely subjective interpretations, while at the same time retaining the role of physics as the only access to asking true questions to the cosmos. The 2010 re-enactment was carried out by the physicist and philosopher Jimena Canales, artist Olafur Eliasson, and philosopher Elie Duran. In an elegant way, the organizing of the event pointed to the productive aspects of organizing conversations—not to settle a controversy once and for all, but to open for new ways of talking, new points of view, and new common languages to emerge. As the organizer Bruno Latour (2011) writes, the conversation was meant to show how the pecking order of disciplines has changed from the twentieth to the twenty-first centuries. The point being that when it comes to fundamental assumptions about time and space, physics no longer necessarily has the higher ground as a 'hard science' making facts that trump the 'softer sciences' of humanities and 'creative endeavors' of artists.

A conversation can make hierarchies of the world visible again and again, but can a conversation leave something in its wake that changes them as well?

1. This text is about curating conversations. It is about the importance of hosting dialogues. The ideas presented come from the curatorial process for the 'Mobile Exhibition' for the Cosmobilities Network's ten-year anniversary conference in November 2014. The exhibition had an open call which five artists and groups responded to. The curatorial process consisted of dialogues between myself and those artists and groups. Ultimately this text is an attempt at conveying comradely concerns about future practices of research in mobilities from the artists and researchers who participated in these dialogues before, during, and after the exhibition.
2. In the conversation at the Centre Pompidou mentioned earlier, the notions of time and space were the themes that stretched out into a conversation. Just as time and space have had central significance for sciences and art, so has mobility. Although perhaps mobility has been given less cosmological, philosophical, and critical theoretical attention (Hannam, Sheller, & Urry, 2006). Movement of bodies and materials, transfer of information and transformation of materials and energy are, like passing of time, integral to the world's becoming (Ingold, 2015, 2007). Not unlike categories of 'space' and 'time,' 'mobility' is a fact of life.
3. Today, the increasing possibilities for mobility is a central axis on which modern civilization turns (Cresswell, 2006; Urry, 2000). From roads,

rails, and sea and air routes to energy grids and fiber optic cables, the infrastructural leaps of civilization have been about increasing movements (Urry, 2007). The logistical optimization of movements of materials (extraction, manufacturing, and supply chain management) have made it possible to convey increasingly complex user experiences and labor regimes (Stiegler, 2013; Kesselring, 2015). Any of the light, speedy, and widespread technologies would not have been possible without a coordination by various experts in physics, chemistry, mathematics, engineering, management, design, and marketing (Sheller, 2014). The list could go on to include the financial industry's acceleration of capital flows, the extractivist corporations' increasingly ambitious underground excavations, or the new media corporations' and hackers' abilities to mediate an increasing flow of images and texts into a growing ecology of media-based public spheres. In other words, globalization as a transdisciplinary project of increased connectivity and circulation is arriving.

4. The distribution of global mobilities seem to be increasingly linked into urban networks. However, urbanization is a process that no longer only describes the transformation of cities. Today what constitutes the urban are layered circuits that makes possible movements of people, materials, information, and energy between cities (Bratton, 2016). Stacked layers of logistical flows are drawing in everything needed to sustain a city and pushing out what is not to and from its peripheral and distant zones of land, ground, water, and air, creating and sustaining global inequalities and spatial segregation corresponding to a more and more fine-grained and sophisticated colonial and capitalist zoning of land into infrastructure spaces (Easterling, 2014). What lies in between city-nodes could be deemed anything from suburban regions or standing rural reserves to exotic destinations or terrains to be overcome or to dump excess waste upon—and thus a necessary urban back end. As globalization is arriving its distribution feels like a giant stack of processes, mainly catering to the logics of what make cities great—often through 'hallmark' events.
5. In many ways these processes correspond to what is now considered an era of the Anthropocene (Ruddiman, 2013; Steffen, Grinevald, Crutzen, & McNeill, 2011; Zalasiewicz et al., 2008). The main idea here being that forces of human labor are intertwining with geological history in ways that are radically changing the future of the planet. Humans have become a remarkably powerful force of nature, embodying more body mass than any other species in the history of the planet, occupying vast amounts of surface area, extinguishing species at the rate of natural disasters, hurling rocks and minerals around at the rate of plate tectonics, and introducing more chemical components to biological evolution and our environments than any other single source (Steffen, Broadgate, Deutsch, Gaffney, & Ludwig, 2015). The Anthropocene is a concept that points to a past we have created transdisciplinarily under the influence of liberal capitalism and modern innovation thinking. It points to a future that demands a series of responses of similar breadth and proportions. What is needed is several leaps of imagination.

6. Here we can return to the Centre Pompidou conversation on a constructive note. Dialogues about theory between disciplines are not only about 'who has the upper hand,' but also about 'who can make the imagination leap.' "It is more important," according to the process philosopher Alfred North Whitehead, "that a proposition be interesting than it be true" (Gaskill & Nocek, 2014: 6). Maybe a time where some grand propositional narratives should be left to fail is also a moment where the role of theoretical abstractions could be rethought. Here the notion of low theory, developed by McKenzie Wark, proposes that abstractions can have an interstitial role, rather than be legislated from above by a 'central theoretical party' (Wark, 2015). The primary task of low theory is to make concepts, narratives, or images that convene abstractions between practices, to let them be tested, appropriated, changed, and recast by different fields of practice. It is a mode of practice that "does not set its own agenda, but detects those emerging in key situations and alerts each field to the agendas of the others" (Wark, 2015: 218). So to rethink what a dialogue can do to theory is maybe less about reversing pecking orders and more about convening between fields of practice—to compose rather than critique, to put it (maybe too) bluntly (Latour, 2010; Stengers, 2005).
7. The role of curatorial practices can be seen as a cunning outwitting of what is considered normal or taken for granted—"to link objects, images, processes, people, locations, and discourses in physical space like an active catalyst, generating twists, turns, and tensions" (Lind, 2010: 57). Often this relies on a number of practitioners that offer their work to proliferate, prolong, or intervene in one or several disciplinary trajectories. I would like to propose a non-institutional telos (ultimate aim) of curatorial practices, where curating moves beyond primarily using the work of artists and researchers to renewing and prolonging what art or academia per se (in itself) could become (Mackay, 2011).
8. Dialogues have been the main reason behind, and the driver and result of my curatorial work around the exhibition that this text is based on. I discovered that talking to people and trying to understand their lives, practices, and the lessons they were learning is what makes me like curating. Dialogues have a resemblance with creative practices in a specific way: They can stretch beyond the questions and assumptions that start them. They don't necessarily take place inside set boundaries, but can traverse presumed or pre-given borders. According to physicist David Bohm (1996), this requires the participants of a dialogue to suspend any expectations that a certain type of exchange is supposed to happen in the conversation (of ideas, knowledge, obligations, credentials, objects, capital, etc.). So often my conversations would start from simple and personal questions like, 'How is your day?' or 'What have you been doing this week?' and slowly continue with more open-ended questions like, 'Where is the tension or excitement in your life right now?' or 'What have you learned from your practice in past months?' This often created quite interesting openings for talking about: the virtual elements of a life and a practice (O'Sullivan, 2006); that which is

not made yet (Obrist, 2014); what work has yet to be done (Stengers, Massumi, & Manning, 2008); or the experiences or forms of knowledge that have not yet been sought out in investigations or experiments (Manning, 2015). In the conversations we imagined the virtual by going backwards and remembering how and why we had been interested in learning in our work and lives. The open-ended dialogues were not only touching on what becomes known or revealed in a process, but also on the assumptions and impulses that initiate and guide how work is done.

9. The 'Mobile Exhibition' became an occasion for continuing those conversations. By organizing a display situation in the junction of the Cosmobilities conference, where participants convened for meals and breaks, the idea was to create a context for encounters and conversations between disciplines. The display of work at the conference were examples of ongoing artistic methodologies of the contributors and the questions that inform or are informed by them. Besides the eventful unplanned encounters and dialogues at the conference, a more formal conversation was organized between myself and the contributors, as a part of both the exhibition and the conference. The idea was to broaden the space for longer term conversations about the curiosity, methods, and dreams involved in artistic and academic practices about mobilities. An audience of about 30 participants followed and participated in the three-hour long conversation, as it moved from each display in the exhibition to the main lecture hall of the conference.
10. What came out of those dialogues between artists (who have all contributed to this book) were things that reached beyond their individual practices. There are many words for activities that sound or feel like interdisciplinary knowledge sharing. Moving ideas between genres of work entail a quite precise account of what is being moved and where to and from. The following three passages are propositions that take steps in the direction of interstitiality—something operating in between or underneath, like the beginning of a 'low theory.' They are interlinked and should be read as inherently important to one another. They do not explain how to deal with globalized forces in the Anthropocene, but are meant to be examples of enabling constraints that could stimulate new types creative engagement with systemic issues.

X:

We Need to Produce Questions That Make Us Organize in New Ways, Rather Than Trying to Answer Questions We Know How to Convene Around Already

Formulating Questions That Operationalize Well

There is an increased urgency in knowing how to ask the right questions. From climate change and mass migration, to increased automation and virtual reality, concerns related to a growing number of real-world events are easily sensed but hard to engage with. What seems to be hard is formulating questions that are worthy of forms of dedication and commitment, that go beyond counting, questioning, and naming. This challenge is not only about how answerable or revealing a question is, but what it demands of those who try to answer it.

What are the current conditions for dedicating efforts to working with questions, beyond just answering them? Do we have the right kinds of institutions to allow researchers and artists to fold processes of actual engagement into their analysis? How can it become easier to move beyond the analytical dialectics of answerability and the aesthetic dialectics of representability?

Y:

We Need to Re-imagine the Relationship Between Work and Employment. Most 'Work to Be Done' Does Not Necessarily Match the Modes of Commitment and Support Offered by Contemporary Research and Art Institutions

Questioning What It Means to Be Employed

The possible relationships enveloped in employment of artists and researchers demand increasing scrutiny. The focus on exchange values of an outcome of a practice seems to have overtaken the interests in the actual processes of exchange embedded in what researchers or artists do. Several modes of work are far more productive and generative in the interior of their making than in the exteriorization of supposed outcomes representing those processes afterwards. A series of questions put forward publicly, ways of moving between contexts or communities, the construction of necessary logistics or infrastructures for a research process can leave more in their wake than the linearity of a text or an object to be put on display.

To rethink the relationships between work, salary, and efficacy is to question how value is attributed to social processes. It is to start working creatively with institutionalized expectations, as something that can be actively changed. Perhaps at this stage, changes cannot be made in the name of cosmopolitan notions of justice or equality, but as a series of concrete attempts to recognize what it is people actually do: not only to make a living, but to live in meaningful ways.

Dedication to work can go far beyond the question of simply being employed. Accepting that a person's life can be an extension of one's practice or the other way around might pose slightly new questions about the assemblage of life-research, life-art-research, or life-research-art. How can research and art support the contemporary ways people and communities are embedded in self-formulated questions, problems, and social changes? How can the movements of dedication and transmissions of collaborative efforts be captured and assessed? What ways of life and mobilities are emerging, as communities of workers move away from protestant work ethics? How can new modes of employment allow more people to charter such immersive journeys? How is it possible to build infrastructures for something we don't yet know what it is?

Z:

We Are Working With Both a Little Bit of White Magic and a Little Bit of Black Magic

White Magic/Black Magic

Any practice is embedded in both creative and destructive forces. Whereas white magic designates a pure and often observatory critical navigation, black magic encompasses the ways it is possible to act structurally and take advantage of systems. The messiness of the world makes it easier to be an observer than to design a practice critically. In other words, carrying a critical attitude outside a discourse and into a real engagement is much easier said than done. There seem to always be a financial surplus to be made, a derivative traded, a carbon molecule emitted, a laborer exploited, a biotope made extinct, an emotional reaction suppressed or oppressed. So without losing sight of pure values and critical reflection, could it be possible to risk these ideals temporarily by putting them to practice?

Taking such risks might give permission to create new subjectivities and invoke new types of agencies in the world. The work of a researcher or an artist could be the incisions, modifications, or augmentations into the economic, technological, and political structures that surround their practice. It might be harder than merely knowing one's ideals to practice the craft of turning capital into broader changes. This could signal a new phase for institutions like the academy, the university, the research network, or the museum. One where the unlearning and unbelonging to the norms and structures that currently guide them lead to a series of new experiments and entrainments in social change.

Could risking norms to discover their usefulness or blightedness be an expanded methodological field? And could the field of mobilities research be one that conveys both white and black magic between practitioners who actively move institutional boundaries?

Conclusions

As this chapter draws to a close, it is important to end with an admonition; the chapter is provisional more than anything. As I would neither consider myself an artist, curator, or proper academic, my hope is not to have drawn a groundbreaking philosophical line, made a significant aesthetic call, or contributed to a grand theory. Rather, by making a set of points alongside of a set of participatory observations in the field of mobilities theory, artistic practices, geological history, and process philosophy, I have tried to convene an occasion for opening up the labor perspective of mobilities research and artistic practices—labor as in how we expect or wish to work in the future, and to what ends those efforts are put in.

The aim of the text has been to challenge researchers and artists to pose questions that unfold creative processes of making or augmenting consequences rather than analyzing to finalize conclusions. Of building affects and affections of consequence to their outsides, their milieus, be they physical, mental, organizational, technological, economic, or environmental. This adds to an ongoing discussion, not necessarily of the subject matter mobilities research engages with, but of its tools for action, augmentation, and manifestation (Merriman, 2014; Sheller & Urry, 2016; Freudendal-Pedersen, Hannam, & Kesselring, 2016).

I have argued for experimentation with unbelonging to institutional frameworks in academia and cultural sectors, and unlearning several habits from their inherent expectations. What this questions is the trajectories of academic institutions (Stiegler, 2015), through the invention of new modes of working and working together (Stiegler, 2017). Questioning how to institute art or research and represent their processes and outcomes is not the same as throwing away solid methods or skills of narration and argumentation that have taken decades to develop. But it might involve risking things that are taken for granted, when it comes to roles of researchers or artists, their means of employment, practices, collaborators, personal engagements, and means of communicating. Unlearning in this respect is about changing what we are familiar with. To learn new ways of working, creating, and living, we might have to forget some of the things that previously seemed central to our ways of life and modes of engagement. Central to this process is to find cross-cutting common grounds, languages, images, and support systems for the turns towards societally oriented presence and actions.

So finally, the idea of this chapter has been to make something useful, by offering a set of challenges that can be taken up in any number of ways by any number and types of practices. Useful in the sense that it could theorize work and put theory to work at the same time.

References

Bohm, D. (1996). *On Dialogue*. Abingdon: Routledge.

Bratton, B. (2016). *The Stack*. Boston: MIT Press.

Cresswell, T. (2006). *On the Move: Mobility in the Modern Western World*. New York and London: Routledge.

Easterling, K. (2014). *Extrastatecraft: The Power of Infrastructure Space*. London and New York: Verso Books.

Freudendal-Pedersen, M., Hannam, K., and Kesselring, K. (2016). Applied Mobilities, Transitions and Opportunities. *Applied Mobilities*, 1, 1.

Gaskill, N., and Nocek, A. J. (2014). *The Lure of Whitehead*. Minneapolis and London: University of Minnesota Press.

Hannam, K., Sheller, M., and Urry, J. (2006). Editorial: Mobilities, Immobilities and Moorings. *Mobilities*, 1(1), 1–22.

Ingold, T. (2007). *Lines: A Brief History*. London and New York: Routledge.

Ingold, T. (2015). *The Life of Lines*. London: Routledge.

Kesselring, S. (2015). Corporate Mobilities Regimes: Mobility, Power and the Socio-geographical Structurations of Mobile Work. *Mobilities*, 10(4), 571–591.

Latour, B. (2010). An Attempt at a 'Compositionist Manifesto'. *New Literary History*, 41, 471–490.
Latour, B. (2011). Some Experiments in Art and Politics. *E-flux Journal*, 23.
Lind, M. (2010). *Selected Writing*. (Brian Kuan Wood, Ed.). Berlin: Sternberg Press.
Macuga, G. (2008). *Sleep of Ulro*. Rotterdam: Veenman Publishers.
Mackay, R. (2011). *The Medium of Contingency*. London: Urbanomic.
Manning, E. (2015). Against Method. In P. Vannini (Ed.), *Non-Representational Methodologies* (pp. 52–71). New York and London: Routledge.
Merriman, P. (2014). Rethinking Mobile Methods. *Mobilities*, 9, 4.
Obrist, H. U. (2014). *Ways of Curating*. London: Penguin Books.
O'Sullivan, S. (2006). *Art Encounters Deleuze and Guattari: Thought Beyond Representation*. New York and London: Palgrave Macmillan.
Ruddiman, W. F. (2013). The Anthropocene. *Annual Review of Earth and Planetary Sciences*, 41(1), 22–38.
Sheller, M. (2014). *Aluminum Dreams: The Making of Light Modernity*. Cambridge: MIT Press.
Sheller, M., and Urry, J. (2016). Mobilizing the New Mobilities Paradigm. *Applied Mobilities*, 1, 1.
Steffen, W., Broadgate, W., Deutsch, L., Gaffney, O., and Ludwig, C. (2015). The Trajectory of the Anthropocene: The Great Acceleration. *The Anthropocene Review*, 2(1), 81–98.
Steffen, W., Grinevald, J., Crutzen, P., and McNeill, J. R. (2011). The Anthropocene: Conceptual and Historical Perspectives. *Philosophical Transactions of the Royal Society A*, 396, 842–867.
Stengers, I. (2005). The Cosmopolitical Proposal. In B. Latour and P. Weibel (Eds.), *Making Things Public: Atmospheres of Democracy* (pp. 994–1003). Karlsruhe: ZKM Publication Program.
Stengers, I., Massumi, B., and Manning, E. (2008). History Through the Middle: Between Macro and Mesopolitics. *Inflexions*, N3.
Stiegler, B. (2013). *What Makes Life Worth Living: On Pharmacology*. Cambridge: Polity Press.
Stiegler, B. (2015). *States of Shock: Stupidity and Knowledge in the 21st Century*. Cambridge: Polity Press.
Stiegler, B. (2017). *Automatic Society 1: The Future of Work*. Cambridge: Polity Press.
Urry, J. (2000). *Sociology Beyond Societies: Mobilities for the Twenty-First* Century. London: Routledge.
Urry, J. (2007). *Mobilities*. Cambridge: Polity Press.
Wark, M. (2015). *Molecular Red: Theory for the Anthropocene*. London and New York: Verso Books.
Zalasiewicz, J., Williams, M., Smith, A., Barry, T. L., Coe, A. L., Bown, P. R., . . . Stone, P. (2008). Are We Now Living in the Anthropocene? *GSA Today*, 18(2), 4–8.

3 Reflective Assemblages

Real and Imagined Mobilities in Locative Media Art

Jen Southern

Art practice as a form of mobilities research has been gathering speed over the past five years through conferences and publications in the European Cosmobilities Network, the Panamerican mobilities network, research centers, and individual practices. This paper will discuss mobilities research and art practice in my own transdisciplinary practice, one developed through my career as an artist and PhD in Sociology. The work I will focus on both produced and used the locative media mobile phone app *Comob Net* to work across the boundaries of art practice, social science research, and speculative app design to research the use of collaborative GPS technologies in everyday mobilities.

To use GPS devices is to be knowingly entangled in a sometimes problematic range of situations and scales that GPS operates on from personal interactions, practical navigation, and ubiquitous surveillance to search and rescue, shipping, and warfare. The uses of GPS in navigation range from globe-spanning international flights to the detailed and local mapping of individual reindeer in the wild, a range of scales that is also evident in locative art projects where GPS has been used to map traces of movement from long Antarctic voyages (Curtis, 2006) to familiar local journeys (Nold, 2004). In its technical operation GPS also uses vast differences of scale from earth-orbiting satellites to atomic clocks. On these different scales, GPS entangles distance and proximity, and its vertiginous scope connects the global and the local, the sky and the ground, and the grid of Geographical Information Systems (GIS) with lived and local experience.

In the transdisciplinary practice that I will describe I use creative research to get closer to these networks through specific located practices of mobility and the real and imagined landscapes of their use. By drawing attention to the operation of GPS technologies that often operate in the background, the work encourages discursive engagement with the creative, communicative, and invasive potentials of tracking technologies.

In this research I look in detail at what it means to be both in touch with other people at-a-distance and on-the-move through locative mobile social networks (De Souza e Silva & Frith, 2010). This is a form of mobility that takes place through physical movement, network technologies, *and*

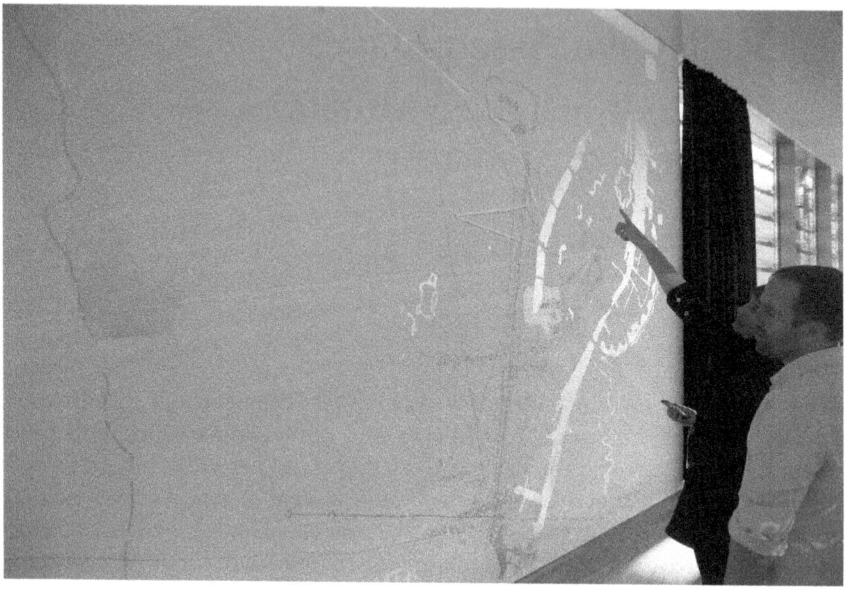

Figure 3.1 *Walking to Work No.3* at Networked Urban Mobilities 2014

Source: Aslak Aamot Kjærulff.

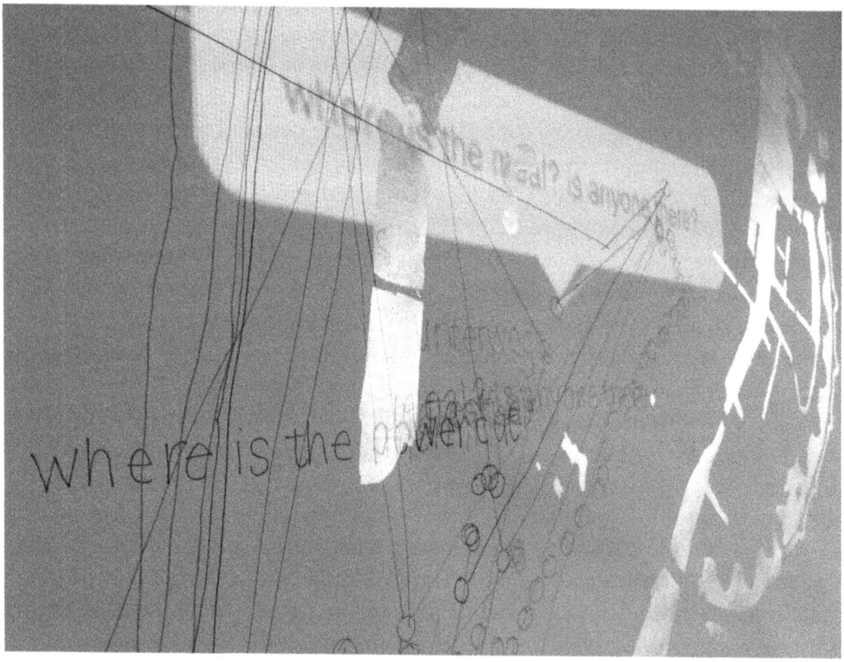

Figure 3.2 Detail of *Walking to Work No.3* at Networked Urban Mobilities 2014

Source: Jen Southern.

imagination. Mobility involves geographical imaginations that "are not simply colorful mental maps confined to the world of ideas. Rather, they are active participants in the world of action" (Cresswell, 2006: 21), and, through the picturing of others in specific locations, contribute to communication through imagining the actions of distant and mobile others.

The urban geographer Edward Soja describes space as real, imagined, and simultaneously 'realandimagined' (1996), a third space that acknowledges the impact that broader spatial, historical, and social contexts have in the operation of everyday situations. The close entanglement of real and imagined can also be found in Donna Haraway's use of figures to understand the intricacies of how things are worldly as "material-semiotic nodes or knots in which diverse bodies and meanings coshape one another" (Haraway, 2008: 4). Through the production of material semiotic figures in artworks and in writing, I aim to involve participants in an embodied aesthetic experience that enrolls the realandimagined in mobilities research through physical movement.

Through description of the *Comob Net* project I will draw out relationships between research, representation, and imagination that are made evident through practice-based research methods and exhibitions as research outcomes. The production of and experimentation with *Comob Net* is simultaneously: *art practice*, in its capacity to involve participants in dialogue and reflection on concepts of everyday imaginative, social, and spatial choreographies; *speculative design*, in its critical reflection on GPS technology and iterative software design; and *mobile sociological method*, in the collaborative discussion and reflection on the ways in which GPS technologies are becoming entangled with everyday mobilities and forms of sociality.

The work has developed from participatory art practice (Bishop, 2006) and mobile methods (Büscher, Urry, & Witchger, 2011), and is part of a wider move towards hybrid art and social science practices including inventive methods that are both creative and that enact new social realities (Lury & Wakeford, 2012), and research in art and mobilities (Witzgall, Vogl, & Kesselring, 2013). These methods are embodied and performative, they enact and produce realities (Law & Urry, 2004), and they have consequences.

Action as Method

The methods used in the *Comob Net* research and *Walking to Work* artwork draw on art, design, and social science through participatory mobile methods in which moving and acting with phenomena enables deeper understandings of situated, spatial, embodied, and technologically augmented interaction. Engaging people in participation in experiences of the everyday in order to explore active forms of knowledge or "doing as knowing" has historically been part of both art practice (Kaprow, 1993: 7) and social science methods (Marres, 2012), and is particularly useful in investigating mobile experiences that are often sensory, mobile, and fleeting (Law &

Urry, 2004; Büscher et al., 2011). Practical engagement in an activity with other people develops different kinds of shared knowledge and more democratic and situated approaches to experience that can be seen in the development of participatory art (Bishop, 2006), designerly ways of knowing (Cross, 2010), and sociomaterial approaches to design-in-use (Suchman, Trigg, & Blomberg, 2002). Walking as a relational practice has also been studied and used as a method in art practice (Myers, 2011) and in sociology through the 'go along' with a research participant that reveals complex and situated relationships to local environments (Kusenbach, 2003). In the field of design 'Future Laboratories' offer opportunities to experiment in real-world contexts to enable things to 'arise concretely' in action (Büscher, Kristensen, & Mogensen, 2007). Similarly 'collective experimentation' is a description suggested by Felt and Wynne in relation to public understanding of science, to describe projects in which "situations emerge or are created which allow [people] to try out things and to learn from them" (Felt & Wynne, 2007: 26).

Through these examples it is clear that doing something together in order to explore and understand it is shared across each of these disciplines as both method and dissemination. Working with methods that are simultaneously *material, live, and relational* allows participants to not only be 'informants' but also to be analysts of their own activities and data through collaborative annotation, and in artworks through live and participatory processes that draw people together for discussion. This combination of and translation between disciplines offers new ways to understand the action that is happening across material, discursive, textual, and performed actions.

Method

The *Comob Net* project drew on each of these methods, through the production and use of a mobile phone app that maps social and spatial relationships by drawing lines that connect participants' locations over a map or satellite image (Figure 3.3). The app and this research is a collaboration with Chris Speed (Edinburgh University) and programmers Jochen Innes and Henrik Ekeus, and has been described in detail elsewhere (Southern & Speed, 2009; Lowry, Southern, & Speed, 2009; Speed & Southern, 2010). The app was developed through a series of nine workshops at arts festivals, exhibitions, and conferences in UK cities in which participants used the app for a series of activities and then collaboratively analyzed how the app had changed their mobility in the city, how they had become attuned to different aspects of the environment, and extended their awareness of other people at a distance. Through these workshops a new sense of what I call 'comobility,' of traveling with other people at a distance through locative media, was both produced and made visible. Having observed that the connective lines in the app engaged participants in one another's activity at a distance, I used this quality in three versions of the work *Walking to Work*: a 78-mile

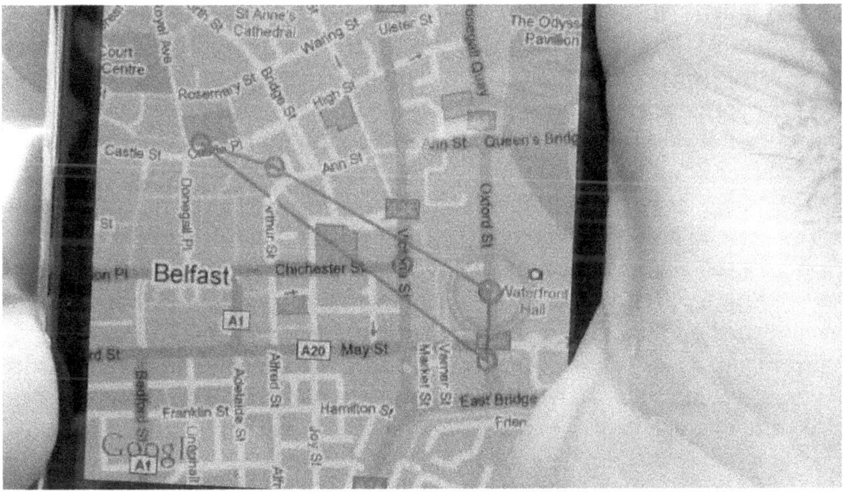

Figure 3.3 The *Comob Net* interface

performative walk over five days during a conference; a two-hour performative walk for a gallery exhibition; and a mapping of conference participants at the Networked Urban Mobilities conference. At the first two events there was a discussion when I arrived at the exhibition space with people who had participated on their phones or followed the action on a projected image.

Comobility

Comobility generates transformations around three forms of presence: *locational presence*, *temporal presence*, and *virtual copresence*. Firstly people are *locationally present*: The blue dot and username displayed in the on-screen satellite image show where people are physically present. Secondly they are *temporally present*: A moving icon is taken to mean that the person it represents is currently connected and there is a sense of a shared 'now.' Thirdly they are *virtually copresent*: Their icons share a spatial relationship on screen, which is reinforced by the lines drawn between them. These three forms of comobility can be used to share time and attention with others at a distance and to vicariously come 'face-to-place' (Urry, 2002), through imagery, discussion, and temporal copresence. A thickness of copresent interactions (Boden & Molotch, 1994: 278) is being augmented by movement patterns made visible in the app, and sequential action can be read in the live and unfolding representation of location, along with commitment to shared time, and the live coordination of mobility that is coproduced at a distance. The availability for proximate interaction of Goffman's response presence (1983: 2) is extended beyond line of sight by live positioning on a map, with both desirable and undesirable consequences. Building on Boden and Molotch's suggestion that interactional detail "engages and entraps" participants (1994: 259) and Licoppe's analysis of compulsion to proximity in locative games (2009), comobile encounters engage us in the sharing of virtual copresence and the possibility of meeting up, but entrap us by making us visible and therefore traceable and accountable (for a more detailed analysis see Southern, 2012).

The imaginative capacities of comobility were revealed in an interview with Liz, who had downloaded *Comob Net* from the App Store and used it to stay in touch and share mobility with her sons (for a more detailed analysis see Southern & Speed, 2015). She described using the combination of live location and satellite imagery to construct a mental image of their physical presence at a distance. The live dot on a map enabled her to construct an image of her sons' activities using either her previous knowledge of their skills and characters (e.g. to imagine one of them working in roadside recovery) or satellite imagery (e.g. of a market in Marrakesh) to vividly imagine them in new situations. She likens this to being copresent with them at a distance. This imaginative travel through *temporal, locational, and virtual* copresence is an important characteristic of comobility.

The real and imagined aspects of comobility were further explored in the art installation *Walking to Work*, which both produces and represents networks. In discussion after the first iteration of *Walking to Work* the lines connecting participants were described as simultaneously geometric abstractions, lines of emotional connection, and evocative of the socio-technical network of human and non-human actors that is necessary to produce comobility: people, satellites, GPS receivers, pylons, and mobile phone networks. This imaginative capacity of the aesthetic experience of the work, which comes about through the experience of using the app, entangles participants in the material semiotics of the technology.

Locative media has been described as a 'hybrid' of digital and local places in which both proximate and distant social interactions are played out together (Gordon & de Souza e Silva, 2011; Willis, 2012; Farman, 2012), but they also combine aerial and eye-level perspectives, enabling the exploration of live embodied action while viewing the world from above through maps or satellite imagery. This aerial perspective is not, however, a God's eye view; it is, as all maps are, a partial and situated perspective (Haraway, 1991). The silver-leaf map of Copenhagen in *Walking to Work No.3* (Figures 3.1, 3.2, and 3.4) that the *Comob* lines are projected onto is a purposefully partial map in two ways. It is not complete, in that accurate locations cannot be specifically identified, only spatial relationships to other participants, roughly located in relation to areas of water. It is also partial because it is rendered in a reflective surface so that how it looks to the viewer depends on where they are standing. Their physical (and metaphorical) orientation to the map and the mapping changes what they see. This rendering of the map as specifically susceptible to an embodied reading,

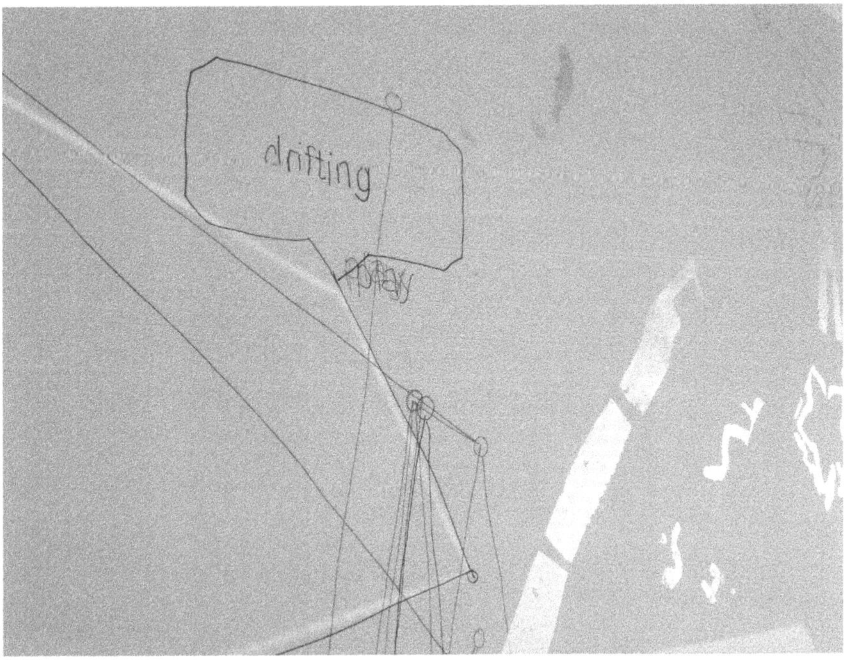

Figure 3.4 Annotation during Networked Urban Mobilities 2014

under construction by participants and in motion, is intended as a rupture in the reading of the map, and is an attempt to problematize an easy reading from above both conceptually and practically. The view from above and in motion is a partial and situated view from somewhere. In making the view from above an embodied and partial one, through its reflective quality, I intend to open the map up for collective annotation and interpretation.

Conclusion

During the Networked Urban Mobilities conference the *Comob Net* app was used by participants to facilitate meeting up for shared events (Figures 3.2), to send short messages about conference papers (Figures 3.5 and 3.6), and to share the mobility of conference delegates. Engagement with the work was, however, not straightforward in this kind of environment; an audience that is largely in one building does not make for interesting spatial tracking. *Walking to Work No.3* aimed to make the technology of shared GPS available for discussion and collective annotation on the map during the conference. In other versions of this work the live-ness of the movement that is visualized intrigues participants, led to speculation about reasons for movement, and was a reason to fix a route by drawing it in order to discuss

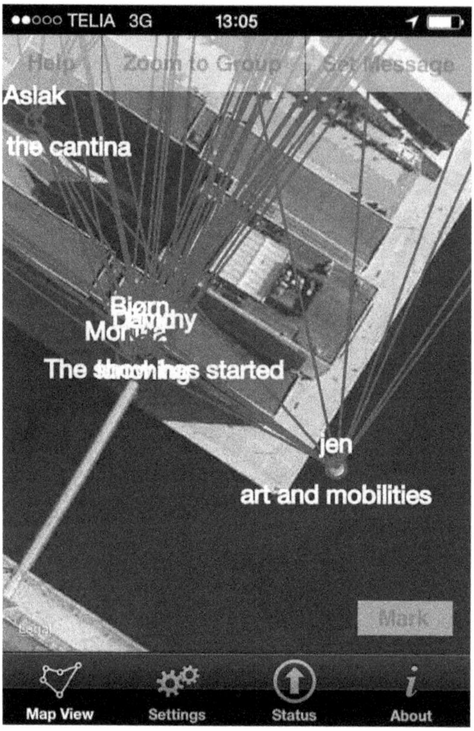

Figure 3.5 Clustering of *Comob Net* users at the conference venue

it later. Within a conference, however, live tracking is often irrelevant when participants are all in the building together and when network data charges for international delegates can prohibit participation. Nevertheless the work still insists on a different form of readership of academic work, one that is actively engaged in an encounter with the phenomena.

The key result of this research and artwork was to produce and make visible the emerging social phenomena of comobility. It not only demonstrated the capacity of collaborative GPS interactions to engage and entrap participants, but was also engaged with new analysis of collaborative GPS use. In addition the project itself was and is also engaged and entrapped by the attendant politics of tracking and surveillance, and by the manufacturing policies of companies who produce the platforms that it operates on. This position is not intrinsic only to this work, but to all uses of location tracking on mobile devices, and as such it is important to consciously engage with devices that entrap.

In pursuing this uncomfortable line of research participants were invited to become part of a reflective assemblage with collaborative GPS tracking through practical experimentation and collective analysis. The work aims

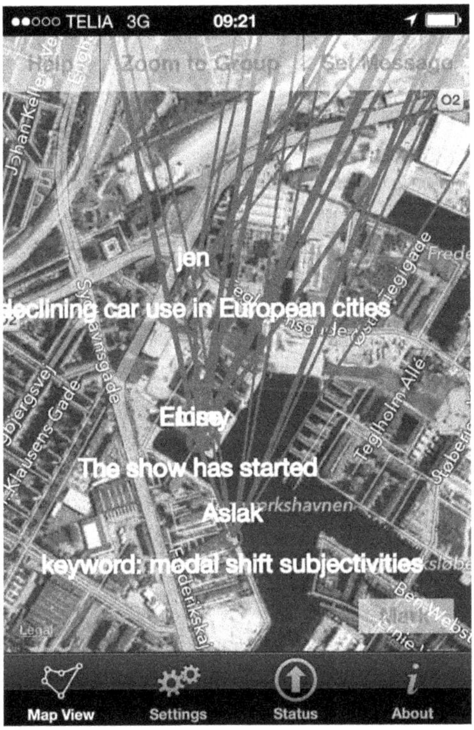

Figure 3.6 Clustering of *Comob Net* users at the conference venue

to reflect on the partiality of maps and views from above through the aesthetics of their representations and to engage in mobilities that are real and imagined, material and discursive, and exist within multiple and relational practices.

References

Bishop, C. (2006). *Participation: Documents of Contemporary Arts*. London: Whitechapel and Cambridge, MA: MIT Press.

Boden, D., and Molotch, H. L. (1994). The Compulsion of Proximity. In R. Fiedland and D. Boden (Eds.), *Nowhere: Space, Time and Modernity* (pp. 257–286). Berkeley: University of California Press.

Büscher, M., Kristensen, M., and Mogensen, P. (2007). Making the Future Palpable: Notes from a Major Incident Future Laboratory. In B. Van de Walle, P. Burghardt and C. Nieuwenhuis (Eds.), *Proceedings of the 4 International ISCRAM Conference*. Delft, The Netherlands, May 2007.

Büscher, M., Urry, J., and Witchger, K. (2011). *Mobile Methods*. Abingdon: Routledge.

Cresswell, T. (2006). *On the Move: Mobility in the Modern Western World.* Abingdon: Routledge.
Cross, N. (2010). *Designerly Ways of Knowing.* London: Springer.
Curtis, L. (2006). *Polar Wandering.* Available at: www.laylacurtis.com/work/display/5-mixed_media Accessed 26 October 2015.
De Souza e Silva, A., and Frith, J. (2010). Locative Mobile Social Networks: Mapping Communication and Location in Urban Spaces. *Mobilities*, 5(4), 485–505.
Farman, J. (2012). *Mobile Interface Theory: Embodied Space and Locative Media.* New York: Routledge.
Felt, U., and Wynne, B. (2007). *Taking European Knowledge Society Seriously.* Report of the Expert Group on Science and Governance to the Science, Economy and Society Directorate, Directorate-General for Research, European Commission. Available at: http://ec.europa.eu/research/science-society/document_library/pdf_06/european-knowledge-society_en.pdf Accessed 26 October 2015.
Goffman, E. (1983). The Interaction Order: American Sociological Association: 1982 Presidential Address. *American Sociological Review*, 48(1), 1–17.
Gordon, E., and de Souza e Silva, A. (2011). *Net Locality: Why Location Matters in a Networked World.* Chichester: Wiley-Blackwell.
Haraway, D. (1991). *Simians, Cyborgs, and Women: The Reinvention of Nature.* New York: Routledge.
Haraway, D. (2008). *When Species Meet.* Minneapolis: University of Minnesota Press.
Kaprow, A. (1993). *Essays on the Blurring of Art and Life.* Berkeley and Los Angeles: University of California Press.
Kusenbach, M. (2003). Street Phenomenology: The Go-Along as Ethnographic Research Tool. *Ethnography*, 4(3), 455–485.
Law, J., and Urry, J. (2004). Enacting the Social. *Economy and Society*, 3(3), 390–410.
Licoppe, C. (2009). Recognizing Mutual 'Proximity' at a Distance: Weaving Together Mobility, Sociality and Technology. *Journal of Pragmatics*, 41(10), 1924–1937.
Lowry, C., Southern, J., and Speed, C. (2009). Modelling the Social in Locative Media: Collaborative GPS. *Second Nature*, 1(2), 135–148.
Lury, C., and Wakeford, N. (2012). *Inventive Methods: The Happening of the Social.* Abingdon: Routledge.
Marres, N. (2012). Experiment: The Experiment in Living. In C. Lury and N. Wakeford (Eds.), *Inventive Methods* (pp. 76–95). Abingdon: Routledge.
Myers, M. (2011). Walking Again Lively: Towards an Ambulant and Conversive Methodology of Performance and Research. *Mobilities*, 6(2), 183–201.
Nold, C. (2004). *Biomapping.* Available at: www.biomapping.net/ Accessed 26 January 2013.
Soja, E. (1996). *Thirdspace: Journeys to Los Angeles and Other Real-and-Imagined Places.* Oxford: Blackwell.
Southern, J. (2012). Comobility: How Proximity and Distance Travel Together in Locative Media. *Canadian Journal of Communication*, 37(1), 75–91.
Southern, J., and Speed, C. (2009). *Watch This Space: From Collective to Collaborative Uses of Locative Media.* E-Science Workshops, Oxford, UK, 2009 5th IEEE International Conference, pp. 188–191.
Southern, J., and Speed, C. (2015). Sharing Occasions at a Distance: The Different Dimensions of Comobility. In V. Hunter (Ed.), *Moving Sites: Investigating*

Site-Specific Dance Performance (pp. 131–146). London: Routledge/Taylor & Francis Group.

Speed, C., & Southern, J. (2010). Handscapes: Reflecting upon the Use of Locative Media to Explore Landscapes. In E. Buhmann, M. Pietsch and E. Kretzler (Eds.), *Proceedings of Digital Landscape Architecture 2010* (pp. 164 – 172), Anhalt University of Applied Sciences. Berlin: Wichmann.

Suchman, L., Trigg, R., and Blomberg J. (2002). Working Artefacts: Ethnomethods of the Prototype. *British Journal of Sociology*, 53(2), 163 179.

Urry, J. (2002). Mobility and Proximity. *Sociology*, 36(2), 255–274.

Willis, K. (2012). Being in Two Places at Once: The Experience of Proximity with Locative Media. In P. Abend, T. Haupts and C. Croos-Mueller (Eds.), *Medialität der Nähe: Situationen—Praktiken—Diskurse* (pp. 177–194). Bielefeld: Transcript.

Witzgall, S., Vogl, G., and Kesselring, S. (Eds.) (2013). *New Mobilities Regimes in Art and Social Sciences*. Aldershot: Ashgate.

4 On Becoming a Parcel

Artistic Interventions as Ways of Knowing Mobile Worlds

Peter Peters

> Modern traveling is not traveling at all; it is merely being sent to a place, and very little different from becoming a parcel.
>
> (John Ruskin, 1819–1900)

The Victorian art critic and essayist John Ruskin famously commented on the experience of train travel in the mid-nineteenth century. Instead of being an art that draws upon individual skills, he argued that traveling by train merely transported people between destinations as if they were parcels (Freeman, 1999: 79). Standardized journeys thus produced "a new kind of boredom" (Löfgren, 2008: 347). Modern travel as it was invented in the days of John Ruskin meant that the work that is necessary when moving from one place to another was increasingly delegated to various actors. These include the traveler who has to plan and improvise along the way, but also governments that have to facilitate and regulate travel, and companies that build infrastructures and offer services to the traveler. The concept of mobilities has helped to open the black box—one might say: the parcel—of 'modern traveling' by describing and analyzing the practices that make up the movement of people, objects, capital, ideas, and information on a global scale, as well as the more local processes of daily transportation, travel, and movement (Urry, 2000; Hannam, Sheller, & Urry, 2006).

In this chapter, I am interested in the role that artistic practices can have in studying these everyday mobilities. More specifically, I want to reflect on some of the epistemological and aesthetic issues that emerge when we consider artistic interventions as ways of knowing mobile worlds (Urry, 2007). In doing so, I draw upon the emerging field of artistic research (Frayling, 1993; Hannula, Suoranta, & Vadén, 2005; Borgdorff, 2006). Here, making art is taken to be a form of doing research and the works of art that result from that research are presented as a form of knowledge (Biggs & Karlsson, 2010). Artworks and performances are thus not only relevant from the perspective of the aesthetic experience, it is argued, but also as epistemic claims. For artistic practice, this development undermines the modernist dichotomy of autonomy and instrumentalism, thus breaking away from the alleged 'otherness' of art as a societal domain that has clear boundaries and can be separated from science (Nowotny, 2010).

I want to investigate how artistic research could contribute to the developing of mobile methods. How can artistic production be seen as a meaningful context to explore mobilities? How can artistic practices open up new ways of understanding and researching the performative ontologies of travel? To answer these questions, I will first briefly explore some general themes and problems related to art as research. Then, I will present two cases from artistic practice and ask how they can be interpreted and analyzed as 'mobile methods' (Büscher, Urry, & Witchger, 2011).

First, I present a sound artwork by the Italian artist and researcher Daniela de Paulis who has extensively worked on mobilities in her artistic practice. In the project *Night Mail*, inspired by the 1936 British documentary showing how mail is transported from London to Glasgow by night train, she made a contemporary sonic documentation of a parcel traveling between these cities. The journey was recorded with an MP3 player inside the package (de Paulis, 2009). Then I will look at a site-specific theater project of the German theater group Rimini Protokoll, who present their work as "the continuous development of the tools of the theater to allow for unusual perspectives on our reality" (Rimini Protokoll, 2016a: 3). In *Cargo Sofia-X* (2006), they adapted a freight truck so that an audience could be seated in its cargo hold. While the truck drivers followed their regular route, the audience looked at actual highways and haulage sites through a large window on the side of the truck. Methodologically, my aim has not been to extensively research both interventions and the artistic practices they originate from in their own right. Rather I frame them as exemplary cases that help to investigate the systemic gestures of transportation and mobilities that stretch out from urban fabrics to connect them over large distances. My focus is not on individual experiences of a journey. By analyzing how artists stage the movements of a parcel and not a person, a qualitative difference is articulated between Flaneuring and inhabiting a system we normally do not know from the inside.

Whereas site-specific arts and performances have been extensively researched and theorized (see Kwon, 2002, for an overview), until recently relatively little scholarly work has been done on the question of what happens when artistic events become mobile (Vogl, Witzgall, & Kesselring, 2013). As Merriman and Webster argue, there is a long history of artists exploring "the aesthetics, sensations and kinaesthetic dimensions of moving through the landscape" (2009: 525), but this relates more to performing the movement as an artistic event, e.g. in art walking (see Collier & Hannam, Chapter 8, this volume; Pink, Hubbard, O'Neill, & Radley, 2010). In this chapter, I aim to show how mobile artistic events such as *Night Mail* and *Cargo Sofia-X* can be seen as apparatuses that visualize or make audible mobile worlds. By analyzing the workings of these apparatuses from both an epistemological and an aesthetic angle, I hope to contribute to the study of the nexus between the arts and mobilities research.

Art as Mobile Research

The debate on art as research combines fundamental philosophical issues of epistemology and methodology with issues of artistic authenticity, as well as institutional and educational strategies. As Borgdorff (2006, 2010) claims, the debate revolves around fundamental questions: When does art practice count as research? What is the object of artistic research and in what ways is it different from the object of scientific research? How can scientific knowledge be distinguished from knowledge generated within artistic practice? Are scientific research methods radically different from artistic methods of research? When answering these questions, it is hard not to be trapped in powerful dualisms: art and science, words and worlds, art practice and art writing, discursive and embodied knowledge, original artworks and their representations. As a practice, art becomes a paragon of unmethodological, autonomous, and intuitive work, while science appears uncreative, methodological, and articulate (Benschop, Peters, & Lemmens, 2014).

How to think beyond such dualisms? Recently, it has been argued that the field of science and technology studies (STS) offers valuable strategies to trace the multiple ontologies of works of art, including works that are presented as forms of research (Yaneva, 2003; Nowotny, 2010). These strategies focus on the 'work of art' as an ongoing endeavor of assembling agencies, rather than as the construction of a singular work that can be (re)presented and categorized in a more or less unproblematic way (Latour & Lowe, 2011; Van Saaze, 2013). To present a given artistic practice as a form of research means ignoring the common-knowledge and the self-understandings of science and the arts, and instead becoming a meticulous follower of the relationality of practice, of what artist-researchers actually do (Becker, 2006; Acord & Denora, 2008; Gramelsberger, 2014; Benschop, Peters, & Lemmens, 2014). Part of this work might be directed to create boundaries between art practice and other societal practices. These boundaries are, however, not pre-given.

In order to be, art has to be performed in multiple and collaborative practices that can be analyzed as art worlds (Becker, 2008). A comparative approach has been developed for the study of the role of mobility in producing mobile worlds (Urry, 2007). Within the new mobility paradigm, journeys are analyzed as networked and performed in and through various mobilities. Practices of travel can thus be thought of as a way of performing places and times (Edensor, 2001; Jóhannesson, 2005). Particular mobilities such as traveling by bus or walking in the city "can become articulated as meaningful activities within different systems and categories of knowledge" and thus as events (Adey, Bissell, Hannam, Merriman, & Sheller, 2014: 15). On the other hand, we could ask how particular events are constituted through "the performative practice of mobile actions" (2014: 15). The emergence of the mobilities paradigm led to a debate on new mobile theories and methods to study these performativities: ways to capture, track, simulate, mimic, and 'go along' with the movements that shape our mobile worlds (Merriman, 2014). Büscher, Urry, and Witchger (2011) have listed methods such as observing the movement of people and objects, participating in patterns of movement, ways of documenting the journey as a time-space event, imagining or even experiencing a journey, as well as art and design interventions such as "the playful appropriation of prototypes of mobile technologies" (2011: 7–13).

The two exemplary artistic interventions that I will discuss below can be understood as a combination of several of these methodological approaches in what Chilton and Leavy (2014; see also Leavy, 2009) have called a merging of social research and the creative arts. There is the observation of movement, or rather the creation of a specific perspective on movement. There is the documentation of the space-time character of the journey, as well as novel ways to create imaginaries of the journey. And in both cases mobile technologies are an integral part of the performance of the journey-as-artistic-intervention. In what follows, I will show how these interventions can be analyzed both from the art worlds perspective as well as the mobile worlds approach.

Night Mail

The Italian artist and researcher Daniela de Paulis has been fascinated by the production of commercial spaces that can make merchandises travel around the world, such as Tetra Pack liquid food containers. "I was interested in containers where you cannot see what is inside. Anonymous parcels traveling. Which I think is symbolic. It can be transposed to people. People traveling in airports are anonymous people" (Daniela de Paulis, personal communication).

De Paulis started to work on her project *Night Mail* in 2009. The title of the project refers to the homonymous 1936 documentary film, directed by Harry Watt and Basil Wright, about a mail train traveling from London to Scotland. The film features a Postal Special train, dedicated to transporting mail. No passengers were on board. It went from London to Glasgow, Edinburgh, and Aberdeen. Exterior shots show the train traveling through the English landscape, collecting new mail at stations along the tracks. Inside the train, the employees of the London, Midland and Scottish Railway (LMS) were filmed while they were sorting the mail en route. W. H. Auden wrote a 'verse commentary' for the film and the score was composed by Benjamin Britten (McLane, 2012).

From this documentary, de Paulis took the idea to send a package from London to Glasgow by mail and record the entire journey with an MP3 player. Being very interested in *musique concrète*, she believes that moments of silence are just as important as situations where a lot is happening. In order to record sounds from inside the traveling package, she had to solve the technical problem that most commercial MP3 devices stop recording after eight hours. With the help of a sound engineer, she was able to customize an MP3 player in such a way that it could record for a week.

> We had to do several experiments with the recorder and also with the box. We had to minimize noises inside the box. Every time someone handles the parcel, a crackling sound would be recorded. We had to find out how to embed the player into a material that absorbed the sound and that reduced the crackling noise as much as possible. It is still there, but it is not very annoying.
>
> (Daniela de Paulis, personal communication)

The MP3 player in the mail package worked. The resulting sound recording is 17 hours and 15 minutes long. De Paulis sees it as a sonic work, but according to her it resembles a book in some ways; because of its long duration most people will not listen to it all at once, but in shorter intervals. At the start of the recording one hears the sounds of people in a post office, of numbers being called to waiting customers, and of the voice of de Paulis when she says that she wants to send the package to Glasgow. The auditory perspective shifts from more active moments when the package is dropped in a bag, transported by a small truck, and handled by mail personnel, to silent passages when it is stored.

> The strongest imagery is when the parcel sits for a few hours in a hangar and you hear people talking in different accents, a Caribbean accent or Indian accent. Your imagination is triggered by these voices. That is the difference between this work and other sound works which are more abstract. It has a movie-like quality. It renders the experience of what we don't see. It leads your attention to what happens in between, just like in the case of containers for merchandise traveling between harbours. We do not really question that this merchandise actually travels for weeks in a row. That is what is most important to our economy, to our lives, remains out of view. What really interests me, is this quality of being unseen—what if we could experience it? Suddenly this journey would have a meaning for us, rather than assuming that things are coming from door to door with a gap in between.
>
> (Daniela de Paulis, personal communication)

Cargo Sofia -X

Rimini Protokoll belongs to the German avant-garde in site-specific theater. The theater group, led by Stefan Kaegi, Helgard Haug, and Daniel Wetzel, developed artistic strategies that produce 'authenticity effects' by designing theatrical situations that generate an impression of close contact with social reality and 'real people.' As Mumford (2013: 153) observes, "in its documentary performances, [Rimini Protokoll] shares aims and methods with a socially engaged ethnography."

In *Cargo Sofia-X* (2006), the group adapted a freight truck so that an audience could be seated in its cargo hold. One of the long sides of the truck was replaced by glass and thus transformed into a 'fourth wall' (see Figure 4.1). From behind the glass, the mobile audience looks at the performance of a reality that normally is not noticed. As if they were parcels, the approximately 45 spectators are driven around by two Bulgarian drivers who follow their regular routes and pass non-places such as roadside fast-food restaurants, cargo handling ramps, warehouses, or border checkpoints (Müller-Schöll, 2008: 66). Rimini Protokoll wanted to stage the cargo handling sites within the host city so as to visualize the abstractions of a globalizing economy. Like many other projects of the group, *Cargo Sofia-X* can be seen as reality theater or 'Theater der Zeit,' as the group calls it (Rimini Protokoll, 2016a). During its two-year season, the production toured to 25 European cities. After that, it was adapted for an Asian context as Cargo Asia (Mumford, 2013).

The projects of Rimini Protokoll have in common that they are creating situations in which the role of the public is extremely ambiguous. Sitting behind the glass wall of the truck, the audience is visible to the public on the streets, that in turn realizes that it is viewed. From the truck, the audience sees a world in which they, in their daily lives as consumers, have their share as actors. This play with the positions of the observer and the observed has been analyzed by Mumford (2013).

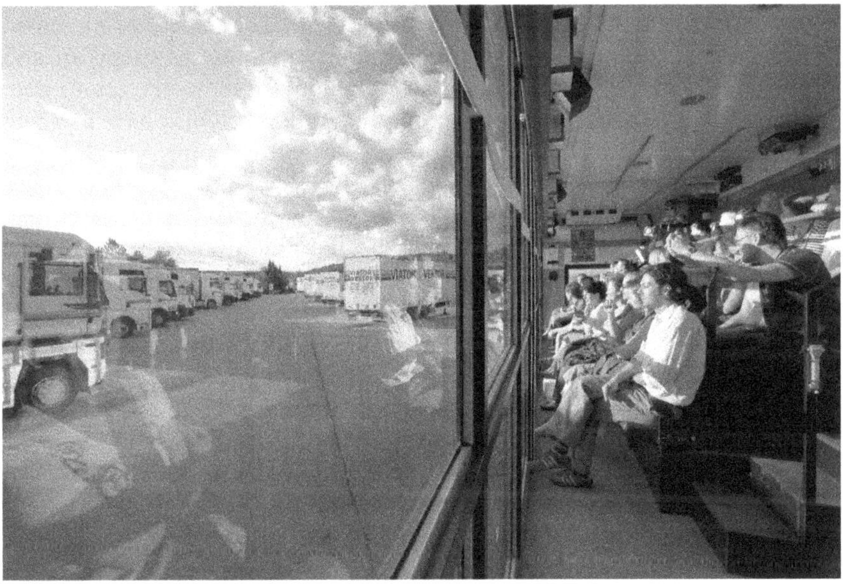

Figure 4.1 Cargo Sofia-X by Rimini Protokoll (Stefan Kaegi)

Source: Photo © Nada Žgank at International festival Mladi levi, Ljubljana 2006.

> Through this play, they questioningly highlight, displace or replace the researcher-artist and spectators' position as the ones who look and know. Situations are created where spectators are made aware of their voyeurism, where they and the theatre practitioners become the observed, and experts become expert observers. Thus, Rimini Protokoll wants to contrive situations that will allow the spectator to experience reality in a new way; to create the feeling for the spectator that all that he has discovered, he has discovered for himself.
>
> (Mumford, 2013: 159–160)

Stefan Kaegi, the artistic director of *Cargo Sofia-X*, used a scientific metaphor to describe the theatrical situation that was produced:

> Where goods used to be stacked in the past, now the audience is sitting and looking from a changed perspective back to their city. Thus the truck serves as an observatory, a theatre probe, a mobile binocular trained at the cities like a microscope.
>
> (Rimini Protokoll, 2016b)

The Machinery of a Theater

How can artistic interventions such as *Night Mail* and *Cargo Sofia-X* open up new ways of knowing the performative ontologies of travel and mobility? One might object that in both cases, research was not the explicit goal of the artists. No epistemic claims were forwarded and no texts were published to reflect on the methods employed in creating them. They were not presented to audiences in the emerging field of artistic research, but to sound art and theater audiences. Yet I would argue that even though these interventions were not intended as research, this does not mean that they can only have meaning in the domain of the arts. As said before, a constructivist approach of science entails ignoring the definitions, boundaries, and differentiations that science itself uses to produce its matters of fact (e.g. Galison, 1987; Star & Griesemer, 1989; Gieryn, 1999). Recently, science and technology researchers have studied boundary crossings between science and other societal practices, such as the arts. In studies of artistic and curatorial practices, a similar approach is taken as in science studies: They do not assume an a priori distinction between scientific facts and works of art (Yaneva, 2003; Latour & Lowe, 2011).

A symmetrical approach means that we compare art and science not at the level of their outcomes, but in terms of the apparatus they put in place to produce these outcomes. In the words of Latour (2004, 2008), we should be interested in the theater machinery that is installed to turn matters of fact—or, to paraphrase, matters of art—into matters of concern:

> A matter of concern is what happens to a matter of fact when you add to it its whole scenography, much like you would do by shifting your attention from the stage to the whole machinery of a theatre. . . .

> Instead of simply being there, matters of fact begin to look different, to render a different sound, they start to move in all directions, they overflow their boundaries, they include a complete set of new actors, they reveal the fragile envelopes in which they are housed. Instead of 'being there whether you like it or not' they still have to be, yes (this is one of the huge differences), *they have to be liked*, appreciated, tasted, experimented upon, prepared, put to the test.
>
> (Latour, 2008: 39)

Can *Night Mail* and *Cargo Sofia-X* be compared to the kind of epistemic machinery that Latour has in mind? I think so. Both have this quality of being a probe. The cleverly designed listening box becomes a means to observe in new ways. It is as if we are able to listen to a journey from inside a parcel in real time and thus experience its temporality, as the temporal structure of the journey is turned into a soundscape. The truck that is turned into a mobile theater creates a perspective in which the everyday life of truck drivers is turned into a storyline that is situated in the networked urban mobilities of our global economy. Speaking about these interventions in terms of matters of concern enables us to explore the work they do as a heterogeneous gathering of agencies. There is no privileged point of view, nor are there hegemonic categories such as the public, autonomy, or even contemporary art to map and navigate these assemblages. Both interventions thus produce situations in which it is possible to like, appreciate, taste, experiment upon, prepare, and put to the test—in short to *know* these hidden urban mobilities by engaging with them in an active and creative way.

The Audience as Co-Creator

The work of Daniela de Paulis and Rimini Protokoll can be analyzed from an epistemological perspective when we borrow insights from science and technology studies. A symmetrical approach, however, should also include art theory and art history. In his seminal essay *Inside the White Cube: The Ideology of the Gallery Space* (1976), Brian O'Doherty argues that a work of art, any work of art, involves a context of perception and experience. This context is never neutral or free from ideology. This is all the more true for the white cube-like rooms in which modernist art is shown, O'Doherty claims. The white cube appears to be functioning as a church where religious truths are displayed, as if untouched by worldly context. O'Doherty analyzes how space and art work together to create a viewer. He does so by making a distinction between the 'Eye' and the 'Spectator' (O'Doherty, 1976: 35ff.). The Eye refers to a disembodied ability to perceive only formal visual properties of a work of art. The Spectator is the embodied though emaciated self that has to reduce his body to the carrier of the eye on entering the white cube. The cube itself is, after all, as removed as possible from the everyday world in which it exists. In the white cube there only remains the analytical, non-situated perception of the work of art.

The critical refection on the role of art institutions and their ideologies was of course mirrored in myriad artistic practices since the 1960s. In the postmodern frame, artworks no longer existed as autonomous and singular objects in a neutral space. In a sense, all art became site-specific, blurring the clear boundaries between the artwork, its mediating institutions, and its audience. O'Doherty's Eye and Spectator were re-embodied in artistic practices that experimented with audiences as co-creator of a work of art or performance. One recent example is the work of artists who were assembled by Nicolas Bourriaud under the heading of 'relational aesthetics,' a term he coined in the 1990s. According to Bourriaud, relational art aims to position itself in the realm of human interactions and social contexts, rather than starting from an independent symbolic space where the aesthetic experience is individualized (Bourriaud, 2002: 14). Not only do artists and audience work together, this collaboration is the artwork.

In similar ways, interventions such as *Night Mail* and *Cargo Sofia-X* question the unequivocality of the audience as consumer of art and seduce it into various forms of co-creation. In *Night Mail*, the artwork can only exist when the listener starts to imagine the journey of the parcel by giving meaning to the sounds it recorded. In the case of Rimini Protokoll, the audience not only watches a world, it also creates it. How exactly? By changing the meaning of social roles, positions, and perspectives in a way that has been analyzed by Hal Foster in his seminal book *The Return of the Real* (1996). According to Foster, avant-garde art at the end of the twentieth century exploits the precarious balance between involvement and distance that is inherent in all ethnographic research. Artists created situations in which the familiar and the foreign vacillate. Precisely the mechanisms that determine what we take for granted and what we experience as strange become the medium of artist as ethnographer (Foster, 1996). It is through a public staging of everyday life in experimental situations that audiences can look at themselves as anthropologists (Schneider & Wright, 2006).

Conclusion

John Ruskin was born in a century when art and science increasingly became separate spheres in society, even though he himself lived them as a continuity. He was an influential art critic who wrote poetry as well as travel guides and essays on topics as varied as geology, ornithology, botany, and political economy. As an artist-researcher-critic, he reflected on the changes in modern travel practices in his time, and he had an influence on tourism discourses that was as important as the practical achievements of Thomas Cook's travel company (Hanley & Walton, 2010). He was interested in what the machinery of the railways did to the perception of time and space of his contemporaries, of their experience of landscapes, and the stories they told about them. Risking the reproach of using a Whiggish anachronism, I would argue that Ruskin developed new mobile methods to reflect on the mobilities of his times in his watercolors, essays, and travel guides.

Arts-based mobile methods cannot claim to render more 'effective,' 'close,' or 'accurate' apprehensions of movements and events, as Merriman (2014) has rightly argued. My interest in *Night Mail* and *Cargo Sofia-X*, however, is not in the accuracy of the facts they do or do not produce, but in the way these artistic interventions stage mobile worlds in novel and compelling ways. They capitalize on art's inherent reflection on the conditions under which it exists as art, in other words on its dependence on theater machinery. As researchers of mobile worlds, we are thus reminded of how our research not only analyses, documents, and informs, but also performs realities and ontologies, and reforms and transforms them through the act of researching (Law, 2004; see also Southern, Chapter 3, this volume).

Analyzing how *Night Mail* and *Cargo Sofia-X* invite the listener and the spectator to become a parcel not only teaches us more about inhabiting networked urban mobilities, but also how these mobilities challenge our ways of researching them. If knowing mobilities assumes 'going along,' then we can ask what theater machinery is put in place to do so and what opportunities are created for the mobility facts to be liked, appreciated, tasted, experimented upon, prepared, put to the test. In the case of arts-based mobile methods, the apparatus to do so is one that creates new site-specificities and vacillating positions between observer and observed, the unusual and the familiar, insider and outsider.

References

Acord, S. K., and Denora, T. (2008). Culture and the Arts: From Art Worlds to Arts-in-action. *The Annuals of the American Academy of Political and Social Sciences*, 619(1), 223–237.

Adey, P., Bissell, D., Hannam, K., Merriman, P., and Sheller, M. (Eds.) (2014). *The Routledge Handbook of Mobilities*. London: Routledge.

Becker, H. S. (2006). The Work Itself. In H. S. Becker, R. R. Faulkner and B. Kirschenblatt-Gimblett (Eds.), *Art from Start to Finish: Jazz, Painting, Writing, and Other Improvisations* (pp. 21–30). Chicago: Chicago University Press.

Becker, H. S. (2008). *Art Worlds*. Berkeley, CA: University of California Press.

Benschop, R., Peters, P., and Lemmens, B. (2014). Artistic Researching: Expositions as Matters of Concern. In M. Schwab and H. Borgdorff (Eds.), *The Exposition of Artistic Research: Publishing Art in Academia* (pp. 34–51). Leiden: Leiden University Press.

Biggs, M., and Karlsson, H. (Eds.) (2010). *The Routledge Companion to Research in the Arts*. London: Routledge.

Borgdorff, H. (2006). *The Debate on Research in the Arts*. Bergen: Bergen National Academy of the Arts.

Borgdorff, H. (2010). The Production of Knowledge in Artistic Research. In M. Biggs and H. Karlsson (Eds.), *The Routledge Companion to Research in the Arts* (pp. 44–63). London: Routledge.

Bourriaud, N. (2002). *Relational Aesthetics*. Dijon: Les presses du réel.

Büscher, M., Urry, J., and Witchger, K. (Eds.) (2011). *Mobile Methods*. London: Routledge.

Chilton, G., and Leavy, P. (2014). Arts-Based Research Practice: Merging Social Research and the Creative Arts. In P. Leavy (Ed.), *Oxford Handbook of Qualitative Research* (pp. 403–422). Oxford: Oxford University Press.

De Paulis, D. (2009). *Night Mail*. Available at: www.danieladepaulis.com/sound.html Accessed 12 January 2016.
Edensor, T. (2001). Performing Tourism, Staging Tourism: (Re)producing Tourist Space and Practice. *Tourist Studies*, 1(1), 59–81.
Foster, H. (1996). *The Return of the Real: The Avant-garde at the End of the Century*. Cambridge, MA: MIT Press.
Frayling, C. (1993). Research in Art and Design. *Royal College of Art*, 1(1), 1–5.
Freeman, M. (1999). *Railways and the Victorian Imagination*. New Haven, CT: Yale University Press.
Galison, Peter. (1987). *How Experiments End*. Chicago: University of Chicago Press.
Gieryn, T. F. (1999). *Cultural Boundaries of Science: Credibility on the Line*. Chicago: University of Chicago Press.
Gramelsberger, G. (2014). A Laboratory View of Art. In M. Schwab and H. Borgdorff (Eds.), *The Exposition of Artistic Research: Publishing Art in Academia* (pp. 102–111). Leiden: Leiden University Press.
Hanley, K., and Walton, J. K. (2010). *Constructing Cultural Tourism: John Ruskin and the Tourist Gaze*. Bristol: Channel View Publications.
Hannam, K., Sheller, M., and Urry, J. (2006). Mobilities, Immobilities and Moorings. *Mobilities*, 1(1), 1–22.
Hannula, M., Suoranta, J., and Vadén, T. (2005). *Artistic Research: Theories, Methods and Practices*. Helsinki, Finland: Academy of Fine Arts Helsinki.
Jóhannesson, G. T. (2005). Tourism Translations: Actor-Network Theory and Tourism Research. *Tourist Studies*, 5(2), 133–150.
Kwon, M. (2002). *One Place After Another: Site-Specific Art and Locational Identity*. Cambridge, MA: MIT Press.
Latour, B. (2004). Why Has Critique Run Out of Steam? From Matters of Fact to Matters of Concern. *Critical Inquiry*, 30, 225–248.
Latour, B. (2008). *What Is the Style of Matters of Concern? Two Lectures in Empirical Philosophy*. Assen: Van Gorcum.
Latour, B., and Lowe, A. (2011). The Migration of the Aura: Or How to Explore the Original Through Its Facsimiles. In T. Bartscherer and R. Coover (Eds.), *Switching Codes: Thinking Through Digital Technology in the Humanities and the Arts* (pp. 275–297). Chicago: Chicago University Press.
Law, J. (2004). *After Method: Mess in Social Science Research*. London: Routledge.
Leavy, P. (2009). *Method Meets Art: Arts-Based Research Practice*. New York: The Guilford Press.
Löfgren, O. (2008). Motion and Emotion: Learning to Be a Railway Traveller. *Mobilities*, 3(3), 331–351.
McLane, B. (2012). *A New History of Documentary Film*. New York: Bloomsbury Academic.
Merriman, P. and Webster, C. (2009). Travel Projects: Landscape, Art, Movement. *Cultural Geographies*, 16, 525–535.
Merriman, P. (2014). Rethinking Mobile Methods. *Mobilities*, 9(2), 167–187.
Müller-Schöll, N. (2008). 'Cargo Sofia-Zollverein' von Rimini Protokoll (Stefan Kaegi). In S. Dinkla and K. Janssen (Eds.), *Paradoxien des Öffentlichen—Die Selbstorganisation des Öffentlichen* (pp. 66–71). Nürnberg: Verlag für moderne Kunst.
Mumford, M. (2013). Rimini Protokoll's Reality Theatre and Intercultural Encounter: Towards an Ethical Art of Partial Proximity. *Contemporary Theatre Review*, 23(2), 153–165.

Nowotny, H. (2010). Foreword. In M. Biggs and H. Karlsson (Eds.), *The Routledge Companion to Research in the Arts* (pp. xvii–xxvi). London: Routledge.

O'Doherty, Brian. (1976). *Inside the White Cube: The Ideology of the Gallery Space.* Berkeley: University of California Press.

Pink, S., Hubbard, P., O'Neill, M., and Radley, A. (2010). Walking Across Disciplines: From Ethnography to Arts Practice. *Visual Studies*, 25(1), 1–7.

Rimini Protokoll. (2016a). Available at: www.rimini-protokoll.de/website/en/about.html Accessed 18 January 2016.

Rimini Protokoll. (2016b). Available at: http://rimini-protokoll.de/website/en/project_108.html Accessed 18 January 2016.

Schneider, A., and Wright, C. (2006). *Contemporary Art and Anthropology* (English edition). Oxford, UK: Berg.

Star, S. L., and Griesemer, J. R. (1989). Institutional Ecology, 'Translations' and Boundary Objects: Amateurs and Professionals in Berkeley's Museum of Vertebrate Zoology, 1907–1939. *Social Studies of Science*, 19(3), 387–420.

Urry, J. (2000). *Sociology Beyond Societies: Mobilities for the Twenty-First Century.* London: Routledge.

Urry, J. (2007). *Mobilities.* Cambridge: Polity Press.

Van Saaze, V. (2013). *Installation Art and the Museum: Presentation and Conservation of Changing Artworks.* Amsterdam: Amsterdam University Press.

Vogl, G., Witzgall, S., and Kesselring, S. (Eds.) (2013). *New Mobilities Regimes in Art and Social Sciences.* Aldershot: Ashgate Publishing, Ltd.

Yaneva, A. (2003). When a Bus Met a Museum: Following Artists, Curators and Workers in an Installation. *Museum and Society*, 1(3), 116–131.

5 Listening to Mobility and Location-Based Media

Verdun Music-Route

Samuel Thulin

This chapter delves into connections between mobilities research and sound studies through an investigation of the artistic practice of the music-route. After considering some important links between sound and mobility, I focus on the Verdun Music-Route, particularly the way it sonically engages ideas of location-based media, participation, and collaboration. The app-based music-route offers an opportunity for participants to modulate their habitual movement through space in order to listen to and perform a mobile composition. Through an analysis of interviews with 12 people who went on the Verdun Music-Route, I examine how these participants perceived both their mediated environment and their role in the artwork. Interviewee responses draw attention to tensions between maintaining and forging senses of place through location-based media, as well as the dynamics of established smartphone gesture repertoires, normative street mobility, and provocations for different ways of moving and interacting with mobile devices. I conclude by emphasizing how sonic and mobile practices are increasingly networked and interdependent, pointing to the importance of sound and listening for mobilities research.

Introduction: Sound and Mobility

We are always hearing mobility, though we may not always be actively listening to it. Insofar as movement of some sort is responsible for all sound, our acoustic worlds may be thought of as composed of interacting forms of mobility. In his seminal sound studies work, *The Tuning of the World*, R. Murray Schafer (1977/1994) suggests that we think of ourselves as simultaneously the audience, the performers, and the composers of the soundscape of the world (p. 205). The simultaneity of these roles both attends to, and happens through, mobility. Sound walks, for instance, have been a core 'mobile method' in acoustic ecology since the inception of the World Soundscape Project in the early 1970s. There is still some way to go, however, in developing a robust mobilities research agenda that considers what might be gained through an intensive dialogue with work in the now burgeoning

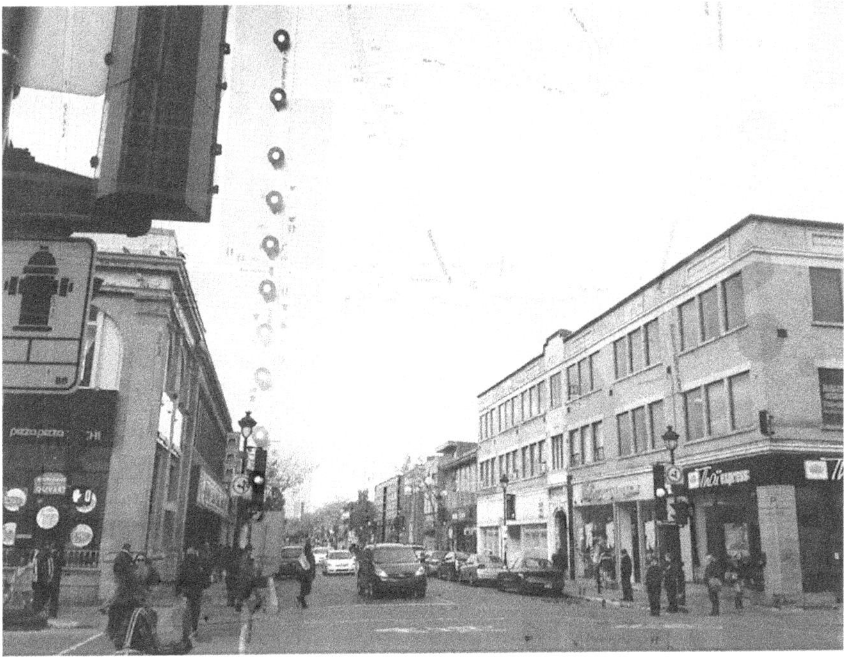

Figure 5.1 Overlay of street view and map view of Wellington Street in Montreal, the location of the Verdun Music-Route

Source: Samuel Thulin + Map tiles © openstreetmap.org contributors.

and multidisciplinary area of sound studies. The recent two-volume *Oxford Handbook of Mobile Music Studies* (Gopinath & Stanyek, 2014b, 2014c) presents a valuable contribution here as the editors offer a "call to scholars of mobility to take music and sound much more seriously" (Gopinath & Stanyek, 2014a: 4). There has also been some earlier work (albeit not a lot) in the social sciences that has explored the sound-mobility connection in sometimes figurative, sometimes literal ways. David Seamon's (1980) 'place ballets,' for instance, suggest an implicit but dynamic musicality and an ensemble of interacting rhythms, compositional and improvisational in nature: "human and material parts unintentionally foster a larger whole with its own special rhythm and character" (p. 163). Henri Lefebvre (1992/2004), in his exposition of rhythmanalysis, suggests that we might learn to " 'listen' to a house, a street, a town, as an audience listens to a symphony" (p. 22), his use of scare quotes suggesting a metaphoricity in the act of listening. Timothy Ingold (2000) meanwhile has drawn a more literal connection in describing the 'taskscape'—the scape in which activities are performed—as accessible through sound, since "what I hear is activity" (p. 199).

Doings, work, comings and goings make sound and tell us something about a place. Listening, then, is not about hearing the static identity of a place, but about attending to and participating in the performance of place, in the sounding of what, following Doreen Massey (2005), we might refer to as constellations of trajectories. Of course, this means thinking not only or simply of one big orchestration out there in a single soundscape, but also of the sounds we carry with us—imagined, remembered, mediated through mobile devices—that influence and overlap with all the other sounds we encounter. There are both the sounds *of* mobility and activity, and the mobile sounds we take with us as we listen to, perform, and compose the places we move through.

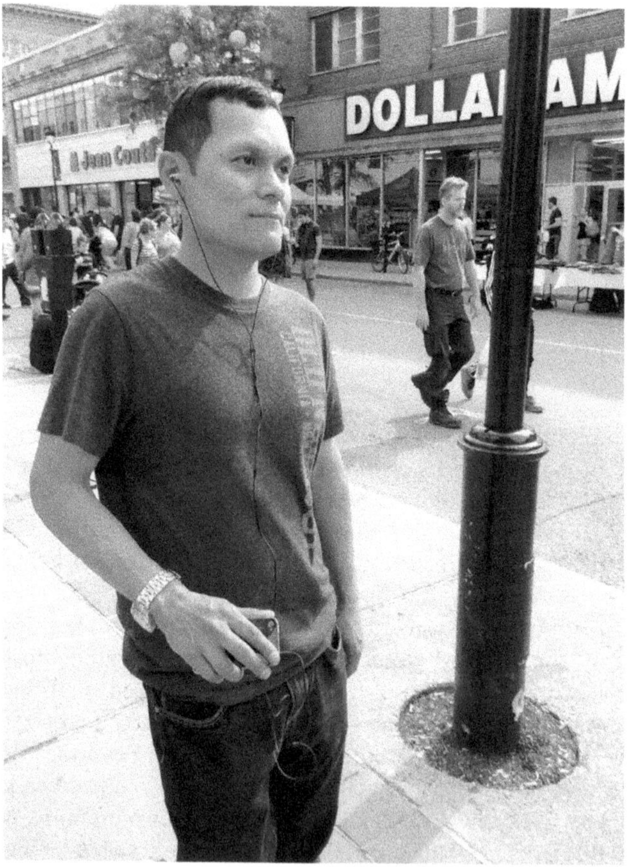

Figure 5.2 Participant performing/interacting with the Verdun Music-Route

Source: Samuel Thulin.

The Verdun Music-Route

The Verdun Music-Route interrogates and intervenes in these sounding mobilities. The basic approach for creating a music-route involves: 1) choosing a route through a place involving any number of forms of transportation; 2) making numerous audio-recordings while traveling along this route at different times on different days; 3) creating a musical composition by editing, transforming, and arranging these audio-recordings; and 4) pairing this musical composition with the selected route by making the piece available for people to listen to on a mobile device while traveling the route. The sounds of mobility, recorded while moving through the place, become mobile mediated sounds and the distinction between what is heard in the headphones and what takes place in the listener's surroundings is blurred. The music-route form draws inspiration from sound walk artists such as Andra McCartney and Hildegard Westerkamp, from the audio walks of Janet Cardiff and Georges Bures Miller, the electrical walks of Christina Kubisch, and the GPS-based sound walks of Teri Rueb. Early music-routes were made available as MP3s that could potentially be listened to anywhere, but were thought of as only being 'complete' when combined with the intended trajectory. The Verdun Music-Route, however, represents a shift both in format and concept, delving more deeply into location-based media.

Unlike a static and linear MP3, the Verdun Music-Route is app-based, geolocated, and interactive.[1] As such, it more thoroughly explores practices of 'mobile mediality,' which Mimi Sheller (2013) describes as a new form of "flexible and mediated spatiality" tied to developments in mobile information and communication technologies (p. 309). Different parts of the composition are geo-tagged to different segments of Wellington Street in the neighborhood of Verdun, Montreal, so that a walker's navigation of physical space coincides with their navigation of the music—they can decide to return to different parts or move through the composition at variable speeds.[2] Participants can also let sounds from their environment in through the microphone to blend with the music, and they can filter these sounds according to musical frequencies by making hand and arm gestures. So each participant hears and takes part in creating a different version of the music-route according to their path and gestures; the audio arising from these interactions is recorded. Between October and December 2013, 12 participants created 12 versions of the Verdun Music-Route, the audio of which is available on Soundcloud: https://soundcloud.com/samuelthulin/sets/verdun-music-route.

The Verdun Music-Route engages ideas of location-based media, participation, and collaboration; in doing so it expands the research-creation agenda of my earlier music-routes. Site-specificity has been central to all music-routes created to date, but the Verdun Music-Route goes a step further by treating location as an element of the interface that allows for direct

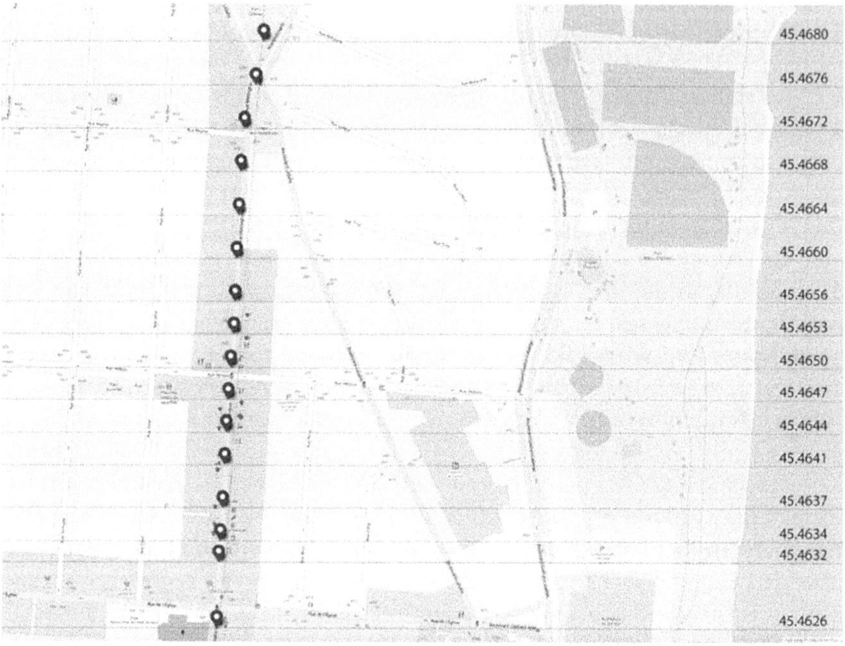

Figure 5.3 Map of the latitudinal zones for the Verdun Music-Route, with Wellington Street marked by pin icons. The music changes each time a participant crosses into a new zone. Since the music-route is geolocated using only latitude, it could be listened to elsewhere around the globe within this swath of latitude

Source: Samuel Thulin + Map tiles © openstreetmap.org contributors.

interaction with the audio. The recording of this interaction means that listeners have audio documents of the way they moved in and through the GPS coordinates that co-ordinate the unfolding of the music. The way the user is involved in the work means that rather than the music-route being only about the combination of music and route, there is an enhanced awareness and play with the relational aesthetics of the situation (Bourriaud, 2002) reaching beyond the fleeting experience of the walk. The work is no longer completely physically site-specific—it can be appreciated in different forms in different places. For instance, participants' recordings are on Soundcloud, where they have been shared with friends and are available to anyone who visits the site. In the project's combination of creative practice and research, the idea of relational aesthetics can be further employed to consider what, borrowing from Douglas Hofstadter (1979/1999, 2007), I have referred to as the strange loops or tangled hierarchies of research-creation—a relationship between the two terms where one seems to fold back into the other, or where whichever term we take to be dominant is simultaneously dominated by the other (Thulin, 2015). In this sense, the artwork is not confined to the combination of the composition and the route, but is diffuse and distributed across every part of the project from recording the sounds, to composing, to contacting participants, to meeting up with participants on-location, to their experience of the route, to the interviews at a local coffeeshop following their walks, to the files being posted and shared online, etc.[3] The differentiation between the 'art parts' and the 'research parts' of the project is purposefully complicated, the aim always being to explore how the vacillation of position, from emphasis on art to emphasis on research within the research-creation loop, can aid in new ways of doing and thinking.

The Interviews

In considering participant responses to the Verdun Music-Route it is important to keep in mind the peculiar dynamics of the situation. Interviewees were aware that I was the maker of the music-route and their responses were inevitably, even if subtly, influenced by this knowledge—for example, they would be less likely to tell me they disliked the work than if I was a 'neutral' third party. Again, this relationship can be framed as part of the particular aesthetics of the work. But in addition, my questions were less about how participants assessed the quality of the music-route and more about how they perceived 1) their mediated environment and 2) their role in the project. The first set of questions revealed a tension between maintaining and forging a sense of place. Unlike the majority of location-based media, the audio of the Verdun Music-Route is not self-evident in its connection to the location—it is not an audio guide or commentary on the place—making its reason for being there questionable. While some interviewees felt that the music fit with the activity and pace of the street, maintaining the place as they understood it, others perceived a conflict between the two, or even a simple disconnection—like two lines of experience flowing but never crossing—preventing them from feeling like it was really meaningful. For others the simple fact that the media was geolocated was enough to make it belong there. These respondents assumed there must be some reason I had joined the place with the music and were curious to figure out what that might be. Here the music-route was likened to a curatorial act in which the juxtaposition of things forged connections rather than maintaining predetermined associations. Participants also made comparisons to music streaming services that provide curated playlists. As one participant put it: "I can feel happy, but I want to know what other people consider happy playlists." This emphasizes the sociality of listening to music, and in the case of the Verdun Music-Route it points to a curiosity around experiencing a neighborhood through someone else's ears.

The dynamics of maintaining and forging also played out in the mobility participants adopted. As the music-route implicitly prompted participants to explore the sidewalk by varying their gait and making out-of-the-ordinary

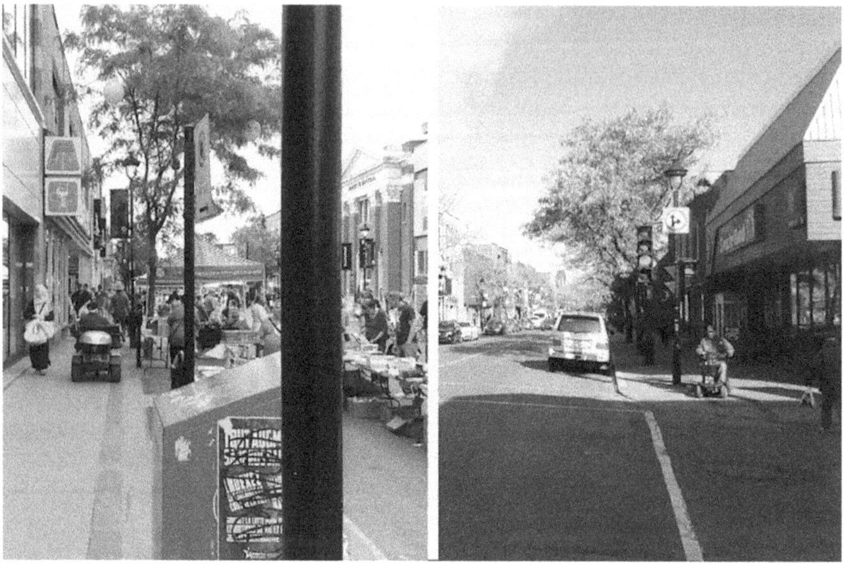

Figure 5.4 Two photos showing the variable activity levels of Wellington Street. Left photo: summer afternoon during a street fair, 2015. Right photo: autumn morning, 2014

Source: Samuel Thulin.

gestures with the phone, it subtly coaxed them from maintaining normative smartphone gesture repertoires (Sawchuk & Thulin, 2016), calling on unusual 'media choreographies' (Bench, 2014) and forging new ways of traversing the street. This activity modulated the street's kinaesthetic field, defined by Jaana Parviainen (2010) as the characteristic motion embedded in a place (p. 320). Michael, one of the participants, offers the example of walking in loops around an intersection because he enjoys the section of the audio that is playing there; while doing so, he imagines someone watching him from their balcony, trying to make sense of his strange behavior. This suggests an irresistible link to Henri Lefebvre's rhythmanalyst for whom the balcony is the ideal liminal position for engaging with the rhythms of the street. As Michael puts it:

> I'd get a kick out of it if someone was just sitting on their balcony watching some loopy guy making some loopy direction decisions. I hope that it does happen. . . . I hope that it makes them, I dunno, have an extra bowl of ice cream during the day or something.

This imagined contagion of minor interventions into the rhythm of the everyday draws attention to what Jason Farman (2012) has referred to as the sensory-inscribed body in physical and digital space, pointing to how the body is simultaneously felt and read. Michael is experiencing his gesture and its effects on the digital media with which he is engaged, *and* he is imagining how others will interpret this gesture. Carrie Noland (2008) argues that gestures have an "energetic charge or 'vitality affect' that overflows the meaning transmitted" (p. xiv). Such excess is experienced first-hand in the body, but it may also be thought of as witnessed by, and transduced in, someone else.

Indeed, gestures exist in different spaces and spill between them, always in excess of any attempt to contain them within a single context. Building on Shuhei Hosokawa's (1984) notion of 'secret theatre' used to describe how Walkman listeners reveal that they have a secret (by listening to headphones) but do not divulge the contents of the secret (what is being listened to), I've likened gestures used with mobile devices to a kind of expanded headphone leakage where the secret overflows into the surrounding area (Thulin, 2013). The secret is no longer a particular piece of music withheld via headphones—it could be any number of ways of interacting with various media, networks, and spaces, manifested through the often oblique gestures these assemblages elicit.

These gestures, in the case of the Verdun Music-Route, influence the course of the audio, resulting in a mix unique to the particular participant. But there are still boundaries to a sense of collaboration and participation in the outcome. Interviewees used a lot of words to describe their perceived roles, including: "one of the people in the band"; "contributor"; "participant"; "collaborator"; "co-composer"; "player"; "person in the orchestra"; "page-turner"; "performer"; "amateur DJ"; "intermediary"; and "worker." Generally, however, they felt they had much less input than I did, since I was the person who made the music-route and determined its meta-rules. In some sense I had built in the possibilities for participation in a way that made contributions seem prescribed. This was the case not only technically, but also in terms of how I framed the project when I described it to participants. One of the first participants noted that she refrained from stopping in at a shop because she thought she was supposed to keep walking. After that, I tried to introduce the project to people by telling them to feel free to walk in any way they wanted, to stop, to go in shops, etc., but this also amounts to a form of direction or suggestion. Many people, understanding their role as research participant as much as visitor to a work of art, were concerned with how they *should* do it. So it was more like following a script or performing a score than collaborating in the creation of that score or an open improvisation. That said, participants did tend to feel more involvement in the experience and feel more ownership over the results when they made more unusual gestures and interventions into their normal mobility. Doing something that stood out, that was out-of-the(ir)-ordinary, that forged rather than maintained ways of moving, was valued more than the inherently unique but taken-for-granted character of each person's habitual style of moving. These acts were simultaneously idiosyncratic contributions to the kinaesthetic field of the sidewalk and, as unusual, had internally felt kinaesthetic power for the participants.

Conclusion

The Verdun Music-Route plays with how people listen to and through mobility. Acknowledging the significance of app-oriented mobile devices for moving and listening means that the soundscape, taskscape, hearable kinaesthetic field, or whatever term one uses for the slice of mobility knowable through audition, must be recognized as constituted by what Benjamin Bratton (2014) refers to in his ruminations on the interfaciality of apps as "a blended co-programming of space and software" (p. 8). Both have their norms and conventions, whether it is the requirement for pace and mobility the sidewalk is 'set' at (Vergunst, 2010: 381), or the differentiations between movements that an app recognizes and endows with meaning; these norms and conventions feed into one another. The body as felt and read, as sensory-inscribed (Farman, 2012), moves within these hybrid spaces and in doing so takes part in convergent fields of sound. As movements and sounds become increasingly networked and interdependent in new ways there is more and more reason to think our sonic lives and our mobile lives together. Not only to think them, but to practice them, and to do this slightly askew, to explore alternative possibilities for how we approach combinations of moving, sounding, and the places we traverse. Lefebvre hypothesizes the rhythmanalyst purposefully intervening in the everyday in the manner of an artist or poet. As something we inevitably perform and compose, the everyday is also something in which we *cannot help but* take part in myriad ways. Lefebvre's approach to listening, whether literal or metaphorical, is apt in that it draws attention to the ongoing inescapability of rhythms, prompting a participative and active listening. An active listening is one in which the listener realizes her role in the simultaneity of listening, performing and composing, and it is such an active listening that may contribute to new rhythms for mobilities research.

Notes

1. The Verdun Music-Route draws its functionality from Reality Jockey Ltd.'s now discontinued app, RjDj, which allowed patches created in the open-source, visual programming language Pure Data to be ported to iOS devices.
2. The Washington, DC, band Bluebrain has created three 'location-aware albums' that work on a similar principle, though they do not draw on the actual sounds of the locations where they are set, and they are more expansive in the variety of routes that can be taken through the music.
3. Considering the specific relationships with each of the participants as part of the artwork adds to the connections with site-specificity and relational aesthetics some resonances with the notion of human-specific art, works that are responsive to individual participants. See: www.cantabile2.dk/en/human-specific/.

References

Bench, H. (2014). Gestural Choreographies: Embodied Disciplines and Digital Media. In S. Gopinath and J. Stanyek (Eds.), *The Oxford Handbook of Mobile Music Studies, Volume 2* (pp. 238–256). Oxford: Oxford University Press.

Bourriaud, N. (2002). *Relational Aesthetics.* (S. Pleasance and F. Woods, Trans.). Dijon: Presses du réel.

Bratton, B. H. (2014). On Apps and Elementary Forms of Interfacial Life: Object, Image, Superimposition. In D. J. Spooky That Subliminal Kid and S. Matviyenko (Eds.), *The Imaginary App* (pp. 1–16). Cambridge, MA: The MIT Press.

Farman, J. (2012). *Mobile Interface Theory: Embodied Space and Locative Media.* New York: Routledge.

Gopinath, S., and Stanyek, J. (2014a). Anytime, Anywhere? An Introduction to the Devices, Markets, and Theories of Mobile Music. In S. Gopinath and J. Stanyek (Eds.), *The Oxford Handbook of Mobile Music Studies, Volume 1* (pp. 1–34). Oxford: Oxford University Press.

Gopinath, S., and Stanyek, J. (Eds.) (2014b). *The Oxford Handbook of Mobile Music Studies, Volume 1.* Oxford: Oxford University Press.

Gopinath, S., and Stanyek, J. (Eds.) (2014c). *The Oxford Handbook of Mobile Music Studies, Volume 2.* Oxford: Oxford University Press.

Hofstadter, D. R. (1979/1999). *Gödel, Escher, Bach: An Eternal Golden Braid* (Twentieth-anniversary edition). New York: Basic Books.

Hofstadter, D. R. (2007). *I Am a Strange Loop.* New York: Basic Books.

Hosokawa, S. (1984). The Walkman Effect. *Popular Music*, 4, 165–180.

Ingold, T. (2000). *The Perception of the Environment: Essays on Livelihood, Dwelling & Skill.* London and New York: Routledge.

Lefebvre, H. (1992/2004). *Rhythmanalysis: Space, Time, and Everyday Life.* (S. Elden and G. Moore, Trans.). London and New York: Continuum.

Massey, D. B. (2005). *For Space.* London: SAGE.

Noland, C. (2008). Introduction. In C. Noland and S. A. Ness (Eds.), *Migrations of Gesture* (pp. ix–xxviii). Minneapolis: University of Minnesota Press.

Parviainen, J. (2010). Choreographing Resistances: Spatial: Kinaesthetic Intelligence and Bodily Knowledge as Political Tools in Activist Work. *Mobilities*, 5(3), 311–329.

Sawchuk, K., and Thulin, S. (2016). More Than Just a Pinpoint: Locative Media and the Chorographic Impulse. *Leonardo Electronic Almanac*, 21(1), 160–176.

Schafer, R. M. (1977/1994). *The Tuning of the World.* Toronto: McClelland and Stewart Limited.

Seamon, D. (1980). Body-Subject, Time-Space Routines, and Place-Ballets. In A. Buttimer and D. Seamon (Eds.), *The Human Experience of Space and Place* (pp. 148–165). London: Croom Helm.

Sheller, M. (2013). Mobile Mediality: Location, Dislocation, Augmentation. In S. Witzgall, G. Vogl and S. Kesselring (Eds.), *New Mobilities Regimes in Arts and Social Sciences* (pp. 309–326). London: Ashgate.

Thulin, S. (2013). Moving Beyond the Auditory Bubble: Apps, Gestures, and Musical Participation. In P. Ross and J. Shtern (Eds.), *TEM 2013: Proceedings of the Technology & Emerging Media Track—Annual Conference of the Canadian Communication Association*, Victoria, June 5–7, 2013.

Thulin, S. (2015). Looping Research-Creation. *Media-N: Journal of the New Media Caucus*, 11(3). Retrieved from http://median.newmediacaucus.org/

Vergunst, J. (2010). Rhythms of Walking: History and Presence in a City Street. *Space and Culture*, 13(4), 376–388.

6 Revealing Roads

The Spectral Sounds of Motorways

David Pinder

> Beneath their glassy, unreadable surfaces, roads hide histories; motorways contain multitudes.
>
> (Joe Moran, *On Roads*, 2009: 17)

This chapter is concerned with ways of revealing roads beyond their 'unreadable surfaces,' with a particular interest in listening. Despite the significance of roads as infrastructures for mobility that profoundly shape environments and lives, they are often overlooked and taken for granted. An ordinary road is "just part of the invisible landscape of the everyday," notes cultural historian Joe Moran, and it is "there to be driven along forgetfully on the way to somewhere else" (2009: 2). He suggests the road is "almost like a separate country, one that remains under-explored not because it is remote and inaccessible but because it is so ubiquitous and familiar" (2009: 8). Roads are typically appraised according to their ability to offer smooth, continuous, and preferably rapid passage. That is particularly the case with highways or motorways that are devoted entirely to vehicles and that, like many modern infrastructures enabling circulation, tend to go largely unnoticed until they fail or become congested, disrupted, or blocked. Such road infrastructures have also received "surprisingly little scholarly attention beyond their historical or utilitarian aspects" (Hvattum, Brenna, Elvebakk, & Kampevold Larsen, 2011a: 2–3). They have certainly attracted less attention in social and cultural studies of mobility than mobile subjects and vehicles, and practices of driving and automobility (Merriman, 2014).

A growing body of work on the social and cultural significance of roads has nevertheless been emerging in a range of disciplines that include geography, sociology, anthropology, archeology, and architectural history. Important to many of these studies is an effort to reveal the complex material-cultural formations of roads beyond their status as engineered objects or as 'non-places,' and to consider the varied ways in which they are inhabited, practiced, represented, and 'placed.'[1] A widely acknowledged challenge is that, as the sculptor Carl Andre once put it in an interview, a "road doesn't reveal itself at any particular point or from any particular point." His widely quoted comment came by way of an analogy for his conception of sculpture, and for the significance of bodily movement for appreciating his linear minimalist constructions with their "infinite point of view." He added: "we don't have a single point of view for a road at all, except a moving one, moving along it" (interview in Tuchman, 1970: 57). In an effort to develop such moving perspectives, many social scientists have advocated methods that enable researchers to immerse themselves in moving worlds so as "to move with and to be moved by subjects." These are intended in different ways to "capture, track, simulate, mimic, parallel and 'go along with' the kinds of moving systems and experiences that seem to characterise the contemporary world" (Büscher, Urry, & Witchger, 2011: 7). Roads may thus appear from the perspective of those driving or cycling along them, for example, with mobile video ethnography deployed as one significant means of 'seeing there' and 'feeling there' (Laurier, 2010; Spinney, 2011).

Methodological debate around these issues in the social sciences is prompting increased interest in practices of art and performance that engage with mobilities. Artistic interventions in this area have much to offer understandings and discourses of mobilities, and in some cases constitute forms of research themselves. The traffic runs the other way, too, with social studies of mobility speaking in significant ways to artistic and performance practices, leading to interdisciplinary or multidisciplinary conversations (for example, Witzgall, Vogl, & Kesselring, 2013; Wilkie, 2015; and other chapters in this volume). These debates connect with wider developing dialogues and collaborations between social-spatial research and creative artistic practice, many of which focus on urban and cultural geographies (Pinder, 2005, 2011; Hawkins, 2011). My chapter seeks to contribute to such discussions by briefly following selected creative engagements with motorways in Britain. As Peter Merriman comments in his important writings on the M1 motorway, a wide range of artists, novelists, film-makers, and song-writers has addressed driving and the distinctive geographies of British motorways, especially in terms of how they are experienced visually from the car (2007: 211–213). The practices I consider here involve, in contrast, slow traveling and listening from a perspective that is on foot.

In the next two sections, I follow selected walkers who listen to motorways by walking along them. In the subsequent section, I then turn to an influential art project that combines sound and walking to address the hidden and contested dimensions of a motorway. I finally offer some remarks in conclusion. Running through the chapter are questions about how motorways might be *heard* differently, and how might this involve reckoning with their ghosts. This entails attuning to the spectral aspects of space, or to what have been termed 'spectro-geographies' (Maddern & Adey, 2008); it also involves taking seriously David Toop's proposition in his book *Sinister Resonance* that "sound is a haunting, a ghost, a presence whose location in space is ambiguous and whose existence in time is transitory." He adds: "The close listener is like a medium who draws out substance from that which is not entirely there. Listening, after all, is always a form of eavesdropping" (2010: xv).

Slow Traveling Motorways

Confronting the apparent mundanity of motorways, a recent volume entitled *In the Company of Ghosts* takes "a headlight view of these often ignored but richly modulated environments and the responses they engender." Drawing together artists, writers, poets, and others, the co-editors Edward Chell and Andrew Taylor seek "to reveal hidden precincts of the motorway rooted in our everyday experience of the car and the motorscapes we inhabit" (Chell, 2012: 14). They focus mainly on British motorways, the first of which opened more than five decades ago.[2] If these constructions have always attracted more muted cultural attention in comparison to some of their international counterparts, notably in the United States and Germany, they nevertheless once promised progress, excitement, glamor, and the democratization of transport (Merriman, 2007). They are associations that have long since faded as fantasies of flow have become increasingly congested, their prospects further darkened by environmental crisis. The cover of the book shows a tower in concrete and glass, resembling an air traffic control tower or space station, with a bridge stretching over a freshly surfaced motorway. A symbol of a new era when it opened as a restaurant at Forton Services near Lancaster, in 1965, the tower is now empty, a heritage site that "presides like a doomed sentinel over a sclerotic M6" (Chell, 2012: 13).[3]

Motorway visions have not died entirely. The book includes proposals by architect Will Alsop for a linear SuperCity, based around the M62 that crosses the north of England from coast to coast. But most contributors reflect on everyday practices and components such as motorway signs, service stations, and the typically overlooked terrains by their sides. The book's main title is taken from a contribution by John Davies, who walked the entire route of the M62, from Hull to his home in Liverpool during the months of September and October 2007. He describes it as "deliberately slow travelling," for the most part conducted within earshot of the route although with detours in towns on the way (Davies, 2012: 19). In his diary of the journey, drawn from a blog composed en route, he conveys the violence, fury, noise, and alienation of the road as well as its beauty and wonder (Davies, 2007). Elsewhere he quotes approvingly the musician and artist Bill Drummond who, in a riposte to the relatively drab reputation of British motorways compared with American freeways, once proclaimed the M62 to be the "greatest motorway ever made" (2002, cited in Davies, 2008: n.p.).[4] If this praise is one inspiration for Davies, then a deeper motivation is his concern to explore the relationships between people and place in a context of increased mobility. Troubled by the displacement and detachment that he sees motorways as exemplifying, he does not turn away but rather moves close. Walking becomes a means to engage with people and places in their entangled ordinariness, to explore the sociability of the road, and to open to its everyday multitudes. The road is a guide through places he scarcely knows, or that he thinks he does but wishes to see afresh.

Davies refers to the journey as one accompanied by ghosts. He is haunted by senses of absences in the presence, and presences in the absence. "Dead roads" are cut by the motorway and later reanimated through wildlife and illicit ways. A village is severed by six lanes of traffic, its high street gone and its "nerve ends" now exposed to the traffic. He follows "the imagined line of the demolished houses through the air in the chasm above the motorway, connecting the old road back to itself, sensing a place which is no longer there" (Davies, 2012: 19–20). He hears "spectral sounds" from unseen traffic, "booming, cracking, clanking" from a bridge above, and "hissing, slashing, whispering" from behind lanes of trees (Davies, 2012: 20; Figure 6.1). Other spirits he senses include those of former residents and radical campaigners as well as those of people who have died on the road, some of them commemorated by floral tributes. And he looks down from bridges on to passing motorists, with their fixed expressions and wheel-grasping hands, and recalls T. S. Eliot's line in *The Waste Land* (in words that are in

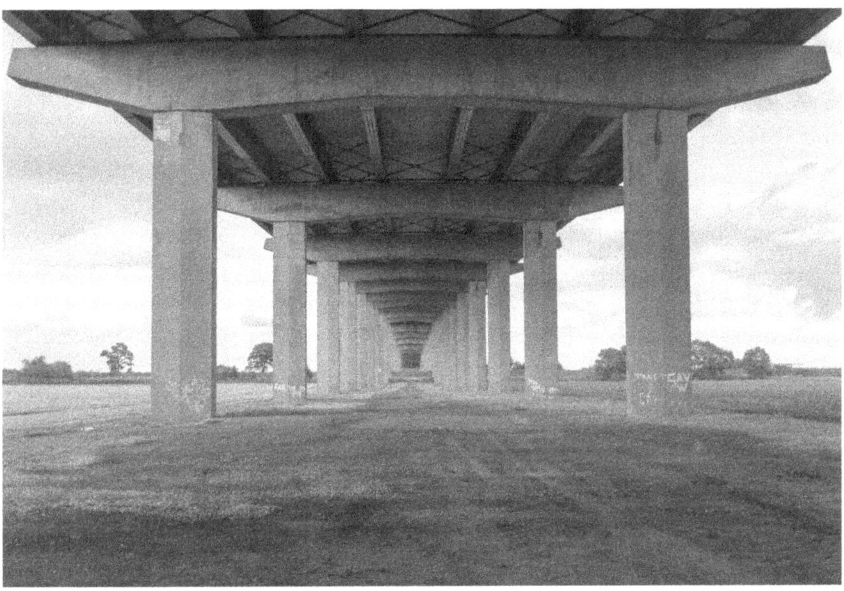

Figure 6.1 "Spectral sounds" under the M62 Ouse bridge at Goole, East Yorkshire, UK: "a few steps into the bridge's shadow remove all vehicular movements from view, and in the sudden stillness the senses open to the booming, cracking, clanking, noises above" (Davies, 2012: 20)

Source: © Phil Hunter/Alamy Photo.

turn drawn from Dante's *Inferno*): "I had not thought death had undone so many" (in Davies, 2012: 23; also 2007: 112).

Against the deadly disconnections of motorway space, Davies counterposes the connections of place. The tenor of his critique is familiar from long-standing traditions of writing on speed and the supposed loss of meaningful place under conditions of modernity (cf. Relph, 1976). In his case it is informed by philosophical and theological writings on place and by less conventional strands of psychogeography.[5] Aspects echo certain critical writings on automobilities that highlight how drivers are insulated from their environments and from each other, how bodies and senses are fragmented and disciplined, and how reciprocal interaction via the eye—what Georg Simmel termed the sociological 'achievement' of the eye—is precluded. From this perspective, argues John Urry, the road can be seen as a world of "ghostly presences moving too fast to know directly or especially to see" (2006: 21). Liquidity triumphs over inhabiting the urban, he adds, and communities become "anonymized flows of faceless ghostly machines" (2006: 22).

As I noted above, other writers have in recent years been wary of sweeping claims about road practices and experiences, and have attended more closely to the highly varied ways of inhabiting roads through driving as well as to the complex placing of motorways (for example Edensor, 2003; Merriman, 2007; Urry, 2007). Davies later acknowledges some of this work for helping him to appreciate other sides to "the mundane roadscape," including the corporeal socialities and imaginative connections that can be fostered through driving (citing Edensor, 2003, in Davies, 2008: n.p.; and Davies, 2012: 23). He is also clear about his own conflicted position as a sometime motorway driver himself who enjoys the opportunities and pleasures that the networks afford. What I want to consider further here, however, concern more the means through which he develops his position, and specifically the adoption of walking and listening as ways of sensuously engaging with the motorway that make manifest aspects of its social spaces and practices, in particular their hauntedness.

Estranging Road Spaces

Walking as addressed above is a deliberate and reflective act rather than a practical means of moving from one point to another. In Davies' case it is not situated as an artistic practice and has more in common with urban pilgrimage, where there is an emphasis on the heightened experience and spiritual understanding to be derived through the prescribed journey. He is particularly interested in its potential for appreciating and reading the everyday. But the practice of exploring roads by tramping on, beside, or across them—in the belief that other meanings might thereby be revealed—is one that has gained currency among a number of contemporary artists, writers, and performers. Among them are several based in the UK that Davies cites as informing his walk, notably Phil Smith, whose 'disrupted walking' performances and prolific writings on the subject have been conducted individually under various aliases as well as in his own name and through the art-research collective Wrights & Sites; Nick Papadimitriou, whose ventures in 'deep topography' take him through overlooked and neglected edgelands around London, which have been sliced by multiple motorways; and the writer Iain Sinclair, who earlier embarked on his own walk around the M25 that became his book *London Orbital*. Addressing the motorway that has encircled London since it opened in 1986, Sinclair resolves that "the best way to comes to terms with this beast was to walk it" (2003: 7). Describing the road as typically more endured than experienced, he believes that trying to fathom it through driving only offers "sensory derangement, diesel-induced hallucinations"; the trick instead is to "step back" and treat it as a "privileged entity," something that walking against its grain enables (2003: 14).

In walking along a route that is designed exclusively for traffic, the pedestrian body is positioned against its motorized counterpart's "iron cage" (Urry, 2006: 20). This can be seen as counterposing practices and perspectives. Among them are public sociality and encounter against the privatized inhabitation of the car; self-generated rhythm against petrol-driven passage; and slow steps against speed. If the slow pace of walking facilitates embodied multi-sensory immersion in place, the rapid passage of the car is to a degree insulated from the external world as sensations are controlled, filtered, and framed by windows and mirrors (Jensen, 2013: 111–115). Inhabitants of cars are shielded from external sounds and able to create their own personalized listening environments in an 'acoustic envelope' (Bull, 2004). There are certainly risks of overdrawing such oppositions, in the process naturalizing walking and exaggerating the enclosure and de-sensing of car inhabitation. Doing so also neglects how these forms of mobility are frequently co-dependent. However, pedestrian incursions into spaces intended for vehicles can, in their strangeness, productively defamiliarize the routines and rhythms of the road. In a recent discussion of the performance practices of various " 'asphalt pedestrians," ' Fiona Wilkie argues that the very incongruity of their movement through road spaces can make motorists "register their own passage"; that is, it can bring into question motoring practices and the socio-spatial processes that enable them, against tendencies for these to be the taken-for-granted norm (2015: 97).

Encountering motorways is a sonic as well as tactile and visual experience. For those on foot—and, of course, for those living and working nearby for whom there is no choice—the road is noise: a juddering of bridges as trucks pass underneath, a modulating roar of combustion engines, a distant hiss, a swish of tires in the rain. At times, for Davies, the motorway is a "screaming carriageway" that assaults the ears, at other times it is an uncanny or "spectral" trace (2007: 25). His route is charted by sound and he listens along the way, especially to his companions and people that he encounters. Sinclair similarly walks not on the motorway but follows in its "acoustic footsteps," which is to say he stays close enough "to catch the hymn of traffic, hot diesel winds" (2003: 16, 77). The noise is "always there," sometimes "barely audible" and a "soothing whisper," while on another occasion it invades sleep in a roadside hotel, "penetrating the double-glazing, sending a shudder through the spine" (2003: 390, 511). In his efforts to recover forgotten and obscured stories from the route, Sinclair listens as well as looks. The road is "many-voiced," he writes (2003: 227), and tuning into voices of those who have gone before is a significant part of his method. He later recalls that, in walking with various companions, they connected with "forgotten ancestors" and in the process: "The noise of the motorway changed from nuisance to a chorus of oracular whispers, prompts, mangled information. Which we had volunteered to transcribe and interpret" (2005: 6–7).

In the next and penultimate section of this chapter, I turn to an artistic project that opened around the same time as Sinclair's account of his walk appeared and that has since been working with the "whispers, prompts, mangled information" of a road space for critical ends. This project is situated in the realms of sound art, where there has long been interest in cultivating ways of listening to urban environments and places, including through forms of site-specific sound and the 'soundscapes' of acoustic ecology (LaBelle, 2015). Recording and replaying the sounds of roads themselves has been one means, an example being an installation about Hume Highway, Australia's most used road that runs between Melbourne and Sydney. This deployed surface noise from passing traffic to explore what is pushed to the side of the road, and became a means of engaging with the ear the road's largely hidden and silenced colonial origins (Lindsay, 2007). The project that I consider next has similar ambitions to reveal aspects of an occluded history of a highway through listening, albeit with reference to more recent events. By bringing pedestrians and the motorway together it plays upon some of the tensions and oppositions discussed above, and it also takes up the theme of haunting. But this work is itself embodied and mobile since it depends upon walking the route and listening to voices and other sounds, layered with those of the current environment. As I now discuss, its background lay in political struggles over road construction that are reflected in its many-voiced form.

Listening to the Displaced: Graeme Miller's *Linked*

A major road building program in the UK was met with strong opposition during the early 1990s when a number of new motorway projects in particular encountered vigorous resistance from activists, residents, and citizens concerned about their environmental and social consequences (Wall, 1999). One of the fiercest and most long-fought of these battles was waged against the creation of a motorway connection—the M11 link road, now called the A12 Eastway—through northeast London (Figure 6.2). This road required

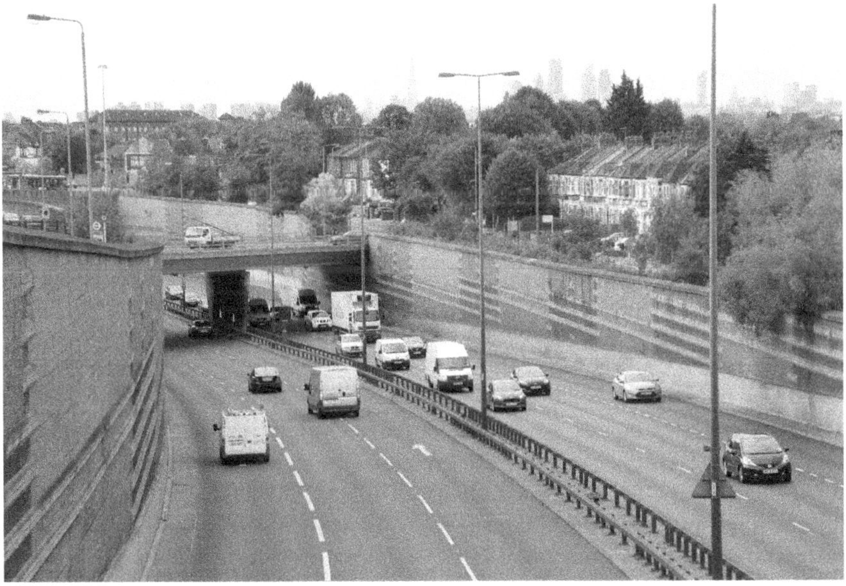

Figure 6.2 The A12 Eastway cutting through northeast London
Source: David Pinder.

the demolition of around 400 terrace houses in the areas of Leyton, Leytonstone, and Wanstead, and the eviction of around 1000 people. Following long efforts to prevent the road from proceeding, construction formally started in 1993 but it was not completed until six years later, having been significantly delayed by the protests. Among those displaced was the artist Graeme Miller, who was thrown out of his home along with his partner and young son only hours before the house was bulldozed. Traumatized by the experience, it was only years later that he resolved to produce a sound walk that engaged with people's stories and memories of the space. Entitled *Linked* (2003), and presented as "a landmark in sound—an invisible artwork—a walk," it was commissioned by the Museum of London and is based on oral histories with some of those who lived, worked, and protested along the route.[6]

Miller's background lies in theater and composition. He typically finds his artistic inspiration and materials in everyday spaces and encounters, often using found or solicited sounds and other fragments to compose a " 'map' of a place or a journey through place." He is particularly interested in the potential of such maps and journeys for "re-awakening memories and re-visioning places" (Phillips, 1998: 103, 104). In the case of *Linked*, the fragments are taken from interviews with former residents. Edited together with music and ambient sounds, they are broadcast via radio transmitters mounted on lampposts close to where the people lived. Each of the 20 transmitters, which are built to transmit continuously for at least a hundred years, is on a loop and can be heard by participants who walk through the area with a receiver and headphones that can be hired for free. The sounds are only discernible close to their source, discovered by moving along a route that often runs near to the road but also passes over bridges and through adjacent streets. Participants are therefore immersed in a shifting and mobile acoustic space. For significant stretches of the route this is dominated by the traffic noise, which is only partly contained by the walls that run alongside the highway, while it is combined with changing sounds of the neighboring streets, and infiltrated by radio transmissions that move in an out of range. This layering invites a re-viewing and re-imagining of the road space.

While some of the voices refer directly to the struggles to stop the road's construction and to being evicted from their homes, the majority tell of everyday life and scenes, in some cases stretching back decades. Invited to imagine the past, the interviewees typically speak in the present tense as if what they are recalling is here and now. Through the medium of sound, long absent lives and spaces are therefore brought into a fragile presence. Narratives are recovered and nurtured. "To make sense of your present world you have to remember history," one of those interviewed recalls his mother as telling him. "And history, by and large, is not great events like Agincourt. It's the myriad experiences of people who have gone before you, which is handed down by word of mouth." This sense of historical meaning and value is shared by *Linked*, which can be seen as "a creative attempt to reinvest the road space with a reminder of lived human experience" (Wilkie, 2015: 94). It is a work of 'civic art' as well as a 'pilgrimage' that involves walking, remembering, and witnessing (Lavery, 2005; Heddon, 2010). The 'walker-listeners' are indeed essential for, as Deirdre Heddon notes, it is only through their embodied movement along this route that the transmitted voices and sounds are received and activated (2010: 38; Figure 6.3). In this sense *Linked* is collaboratively created as participants collectively reweave space through tuning in to the testimonies—this memory, this event, this place—and enacting their own spatial practices, trajectories, and narratives. In its low-key engagement with the politics of everyday urban space, the work therefore resists the erasure of memory that is instantiated by the road and that is daily reinforced as users so easily pass along it, oblivious to its contested histories and geographies. Miller's own hopes are that it "will connect with triggered rememberings, counter-stories, gross omissions and alternative versions of the same event" and, in the process, "may renew the narrative tissue of the neighbourhood" (2003: 2).

The surface of the road does not readily give up its secrets. Looking at the streaming six-lane channel, it is difficult to imagine the streets, homes, gardens, and lives that were once here. Also easily forgotten is the strength of the protest action, first against the building of the road and then, when that immediate struggle seemed lost, against car culture more generally and against the capitalist social structures that sustain it. Miller sees the

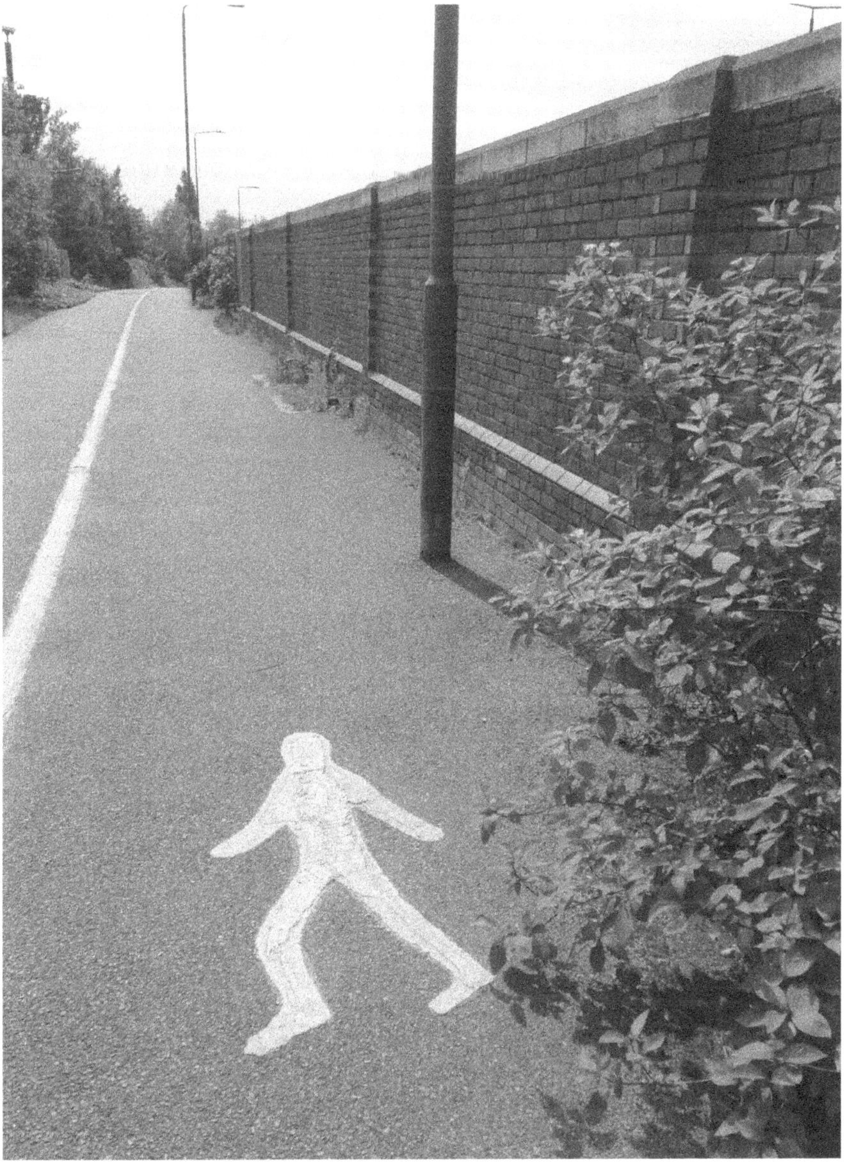

Figure 6.3 Walking beside the highway

Source: David Pinder.

motorway as a sterilizing force that repels stories, that kills the "ecology of narrative." He comments more broadly on "the no-time, no-place, nothing-to-tell feel" of the side of motorways (2006: 108). Through *Linked* he seeks to reclaim that ecology through encouraging attention. This is the kind of attention that is usually all too difficult to muster in daily life, under pressures of time and the need to get from one point to another as quickly and efficiently as possible, and it is directed towards what is here as well as what is no longer. The form of his sound walk is important in this regard, most obviously in the slowness of its pace relative to the speeding traffic. The sound loops further slow down listening by fragmenting speech and through repetition, samples, gaps, silences, and music.

Particularly striking are the disjunctions between what is heard through the headphones—often detailed evocations of people and places, being spoken in the present tense—and what is seen now: a highway, a car park, a narrow strip of green, a noise barrier wall. Places that have physically gone are imaginatively reinscribed in the landscape. This is similar to how, on encountering the village severed by the M62 and with its nerve endings exposed, Davies traces a line of demolished houses through the air and tries in his mind to reconnect the old road. In the case of *Linked*, however, hearing fragments from places that have been so radically transformed is disorienting. It dislocates both space and time, working in a mode that I have discussed further elsewhere as *interruptive* (Pinder, 2016; see also Lavery, 2005). Elements of the past whisper, prompt, and shadow participants in ways that lend the route a haunted feel, and that interrupt the flow of perception and experience. The process is distinct from the "ghosting of the sidewalk" more ordinarily enacted by wired-up walkers listening to MP3 players or iPhones (LaBelle, 2010: 98). This is not least in the way that the source of the sound is now not self-directed and instead becomes the environment itself, or rather the continuously broadcasting transmitters. Audio-recording and radio waves allow voices of the evicted to return, unsettling clear distinctions between absence and presence along with orders of the past, present, and future. They disturb the road's apparent mundanity and inevitability; close listening reveals multitudes.

Conclusions

Roads and specifically motorways have been discussed in this chapter as crucial infrastructures for networked mobilities that, in part due to their familiarity, are frequently overlooked. Earlier I noted how much recent research on mobilities and their spaces, including that on roads, has emphasized the value of mobilizing inquiries and developing means of 'being there,' 'feeling there,' or 'seeing there' through tracking, shadowing, and going along with moving systems and experiences. Researchers have much to gain, it is asserted, if "they capture, track, simulate, mimic and shadow the many and interdependent forms of intermittent movement of people, images, information and objects" (Büscher et al., 2011: 7; also Sheller & Urry, 2006). That might all be the case but the moving perspectives that I have considered in this chapter take a contrary path, one that is conducted against the motorway's grain and is more attuned to absence and loss. Adopting practices and paces that are out of step with dominant rhythms, they involve listening to sounds and to the voices of people along the way. Part of their aim is to defamiliarize the road and to let aspects of its histories and geographies shadow the pedestrian, something that may be understood as a form of haunting that undermines boundaries between absence and presence, and between past, present, and future.

These themes are vividly present in Miller's *Linked*, a time-based work of civic art that takes on different meanings as the urban spaces and social conditions through which it moves themselves change. As I have noted, the form of the work is particularly significant, entailing an embodied encounter with the road space along with the slowing of perception relative to the onrushing traffic. Its deployment of analog technology, which involves tuning in to radio waves, is further taking on new significance in relation to rapidly developing urban digitalization and 'smart' technologies. None of the discussion in this chapter is meant to imply that such practices might be presented as models for social science research; rather, my point is to underline the critical potential of developing dialogues between these fields and disciplines, and specifically between art and mobility studies, for further expanding imaginative and critical engagements with mobility practices and infrastructures. Artistic practice becomes here a means of slowing down matters, of productively making strange ordinary road spaces, and of interrupting senses of the given and inevitable so that those spaces might be perceived, understood, and acted upon differently.

Linked more specifically poses questions that are directly felt at the bodily level about those who have been displaced to enable traffic flow through this area. The apparent solidity and inevitability of the infrastructure are unsettled through this attention to the area's pasts, and also to its potential futures. As a long-term work it asks, what will have become of this area when participants walk through it in ten years', thirty years', or seventy years' time, presuming the transmitters are still functioning then as planned? Might the transmitters even outlast the motorway? Towards the end of his walk along the M62, after being "pummelled in the tumult of motorway ghosts speeding past me," Davies asks a similar question. He has a vision of a day when the motorways may themselves become "dead roads, mossed over, and people will fantasise about the roaring violent convoys that once used them." He notes that when those future people look back, "We will be ghosts to them" (2007: 113).

Notes

1. Examples include Dalakoglou and Harvey (2012); Hvattum et al. (2011b); Merriman (2007); Moran (2009). The journal *Transfers* has also recently published two special sections devoted to 'Roads' (2012) and 'Representing Roads' (2015). On the motorway as an archetype of 'non-place,' see Marc Augé (2009 [1992]).
2. The first major motorway in Britain—the M1—opened in part in 1959, and there are now approximately 3700 kilometers of motorway stretching across the country.

3. The Pennine Tower Restaurant closed in 1989, and the structure was formally listed by English Heritage on 15 October 2012. An illustrated history is at http://fortonservices.webs.com.
4. Drummond writes: "Chuck Berry can keep his Route 66, Kerouac his two-lane black top, Paul Simon his New Jersey Turnpike, Billy Bragg his A13. Give me the M62. Driving it east to west is always best, especially at the close of the day into the setting sun" (2002, cited in Davies, 2008: n.p.).
5. Davies cites among others Edward Casey and Walter Brueggemann, along with the significance of liberation theologies for developing a notion of urban theology. He also remarks on being encouraged towards a more open, global sense of place through encounters with Doreen Massey, in both text and person during his journey (Davies, 2007: 10, 13, 90–91).
6. The project website is at: www.linkedm11.net. A team of five interviewers conducted the oral histories and the recordings are retained by the Museum of London. For background, see Butler and Miller (2005).

References

Augé, M. (2009 [1992]). *Non-places: An Introduction to Super-Modernity*. (J. Howe, Trans.). Second edition. London: Verso.

Bull, M. (2004). Automobility and the Power of Sound. *Theory, Culture and Society*, 21(4–5), 243–259.

Büscher, M., Urry, J., and Witchger, K. (2011). Introduction: Mobile Methods. In M. Büscher, J. Urry and K. Witchger (Eds.), *Mobile Methods* (pp. 1–19). London and New York: Routledge.

Butler, T., and Miller, G. (2005). Linked: A Landmark in Sound, a Public Walk of Art. *Cultural Geographies*, 12, 77–88.

Chell, E. (2012). Foreword. In A. Corkish, with E. Chell and A. Taylor (Eds.), *In the Company of Ghosts: The Poetics of the Motorway* (pp. 13–16). Liverpool: erbacce-press.

Dalakoglou, D., and Harvey, P. (Eds.) (2012). Roads and Anthropology. Special Issue of *Mobilities*, 7(4), 459–586.

Davies, J. (2007). *Walking the M62*. Raleigh: Lulu Press.

Davies, J. (2008). *Walking with the Psychogeographers*. Talk given at Greenbelt Festival, Cheltenham, UK, 25 August. Available at: www.johndavies.org/sermons/talk-gb-08-08-25.html Accessed May 2016.

Davies, J. (2012). M62: In the Company of Ghosts. In A. Corkish, with E. Chell and A. Taylor (Eds.), *In the Company of Ghosts: The Poetics of the Motorway* (pp. 19–24). Liverpool: erbacce-press.

Drummond, W. (2002). *How to Be an Artist*. London: Penkiln Burn.

Edensor, T. (2003). M6: Junction 19–16: Defamiliarising the Mundane Landscape. *Space & Culture*, 6(2), 151–168.

Hawkins, H. (2011). Dialogues and Doings: Sketching the Relationships Between Geography and Art. *Geography Compass*, 5(7), 464–478.

Heddon, D. (2010). The Horizon of Sound: Soliciting the Earwitness. *Performance Research*, 15(3), 36–42.

Hvattum, M., Brenna, B., Elvebakk, B., and Kampevold Larsen, J. (2011a). Introduction: Routes, Roads and Landscapes. In M. Hvattum, B. Brenna, B. Elvebakk and J. Kampevold Larsen (Eds.), *Routes, Roads and Landscapes* (pp. 1–9). Aldershot: Ashgate.

Hvattum, M., Brenna, B., Elvebakk, B., and Kampevold Larsen, J. (Eds.) (2011b). *Routes, Roads and Landscapes*. Aldershot: Ashgate.
Jensen, O. B. (2013). *Staging Mobilities*. London: Routledge.
LaBelle, B. (2010). *Acoustic Territories: Sound Culture and Everyday Life*. London: Continuum.
LaBelle, B. (2015). *Background Noise: Perspectives on Sound Art*. London: Bloomsbury. Second edition.
Laurier, E. (2010). Being There/Seeing There: Recording and Analysing Life in the Car. In B. Fincham, M. McGuiness and L. Murray (Eds.), *Mobile Methodologies* (pp. 103–117). Aldershot: Palgrave Macmillan.
Lavery, C. (2005). The Pepys of London E11: Graeme Miller and the Politics of Linked. *New Theatre Quarterly*, 21(2), 148–160.
Lindsay, K. (2007). Enterprise and Encounter: A Colonial Soundtrack of the Hume Highway. In R. Brandt, M. Duffy and D. MacKinnon (Eds.), *Hearing Places: Sound, Place, Time and Culture* (pp. 402–415). Newcastle: Cambridge Scholars Publishing.
Maddern, J., and Adey, P. (Eds.) (2008). Spectro-Geographies. Theme issue of *Cultural Geographies*, 15(3), 291–393.
Merriman, P. (2007). *Driving Spaces: A Cultural-Historical Geography of England's M1 Motorway*. Oxford: Blackwell.
Merriman, P. (2014). Roads. In P. Adey, D. Bissell, K. Hannam, P. Merriman and M. Sheller (Eds.), *The Routledge Handbook on Mobilities* (pp. 196–204). London: Routledge.
Miller, G. (2003). *LINKED: A Landmark in Sound, an Invisible Artwork, a Walk*. London: Arts Admin.
Miller, G. (2006). Through the Wrong End of the Telescope. In L. Hill and H. Paris (Eds.), *Performance and Place* (pp. 104–112). Basingstoke: Palgrave Macmillan.
Moran, J. (2009). *On Roads: A Hidden History*. London: Profile.
Phillips, A. (1998). Borderland Practice: The Work of Graeme Miller. In N. Childs and J. Walwin (Eds.), *A Split Second of Paradise: Live Art, Installation and Performance* (pp. 102–16). London: Rivers Oram Press.
Pinder, D. (Ed.) (2005). Arts of Urban Exploration. Theme issue of *Cultural Geographies*, 12(4), 383–526.
Pinder, D. (2011). Errant Paths: The Poetics and Politics of Walking. *Environment and Planning D: Society and Space*, 29, 672–692.
Pinder, D. (2016). Sound, Memory and Interruption: Ghosts of London 's M11 Link Road. In S. Jordan and C. Lindner (Eds.), *Cities Interrupted: Visual Culture and Urban Space* (pp. 65–83). London: Bloomsbury.
Relph, E. (1976). *Place and Placelessness*. London: Pion.
Sheller, M. and Urry, J. (2006). The New Mobilities Paradigm. *Environment and Planning A*, 38(2), 207–226.
Sinclair, I. (2003). *London Orbital: A Walk Around the M25*. London: Penguin.
Sinclair, I. (2005). *Edge of the Orison: In the Traces of John Clare's 'Journey out of Essex'*. London: Hamish Hamilton.
Spinney, J. (2011). A Chance to Catch Breath: Using Mobile Video Ethnography in Cycling Research. *Mobilities*, 6(2), 161–182.
Toop, David. (2010). *Sinister Resonance: The Mediumship of the Listener*. London and New York: Continuum.
Tuchman, P. (1970). An Interview with Carl Andre. *Artforum*, 8(10), 57.

Urry, J. (2006). Inhabiting the Car. In S. Böhm, C. Jones, C. Land and M. Paterson (Eds.), *Against Automobility* (pp. 17–31). Oxford: Blackwell.

Urry, J. (2007). *Mobilities*. Cambridge: Polity.

Wall, D. (1999). *Earth First! and the Anti-roads Movement: Radical Environmentalism and Comparative Social Movements*. London: Routledge.

Wilkie, F. (2015). *Performance, Transport and Mobility: Making Passage*. Basingstoke: Palgrave Macmillan.

Witzgall, S., Vogl, G., and Kesselring, S. (Eds.) (2013). *New Mobilities Regimes in Art and Social Sciences*. Aldershot: Ashgate.

7 Developing *Colony*

Objects for Investigation; Spaces for Conversation

Nikki Pugh

Experiencing the World Differently

For an hour or so in October 2011 I walked around the Lower East Side of Manhattan with lengths of thin wooden dowelling attached to each of the fingertips of my right hand (Figure 7.1). I discovered I was able to reach up and touch subway signs, feel opposite sides of tree trunks at the same time, and also explore the sound of railings and other street furniture as I walked past. This simple modification to my interface with the world enabled me to experience my surroundings in powerfully different ways to which I was accustomed.

I was taking part in the 'Urban Sensation Transformer' workshop run by Kaja Kühl as part of the BMW Guggenheim Lab. A few days earlier I had been at the 'Mapping the Distributed Self' session run by creative design consultancy Spurse who had—although I didn't fully realize it then—introduced me to the concept of assemblages. Spurse also re-introduced me to the idea that the conditions under which you experience the world affect what face of the world shows up: the example they gave was that, experienced at everyday speeds, water is a liquid but that if you dive into it from a great height, what you effectively encounter when you hit the surface is a solid.

Thus I began thinking about how the objects I make and the contexts within which they are experienced can together be seen as tools which, through their use, have the potential to reveal differing aspects of the world.

Heavy Objects and the Weight of Convention

After returning to the UK, I produced *Possibility Probe (Heavy Object and Built Environment)*: large wooden tubes carried by participants around the city of Birmingham, UK (Figure 7.2). Real-time processing of Global Positioning System (GPS) data gave a measure of positional error induced by reflection of the time-coded radio signals from buildings and other elements of the built environment, a phenomenon more formally known as 'multipath error.' A simple electro-mechanical system inside the tubes enabled them to produce a tapping noise in response to the measure of error; the more built-up their immediate surroundings were, the more noise they

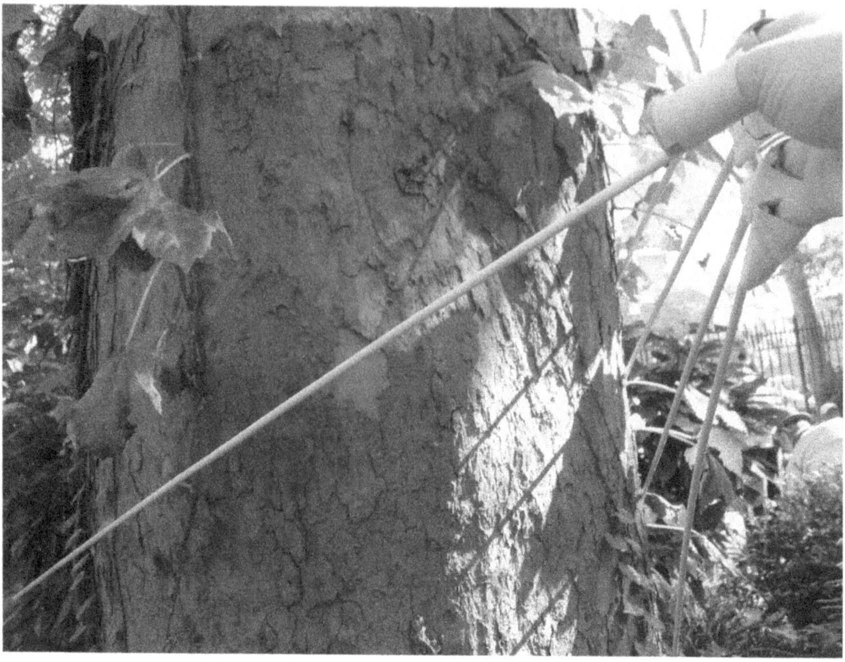

Figure 7.1 Exploring a mediated and expanded sense of touch

Source: Nikki Pugh.

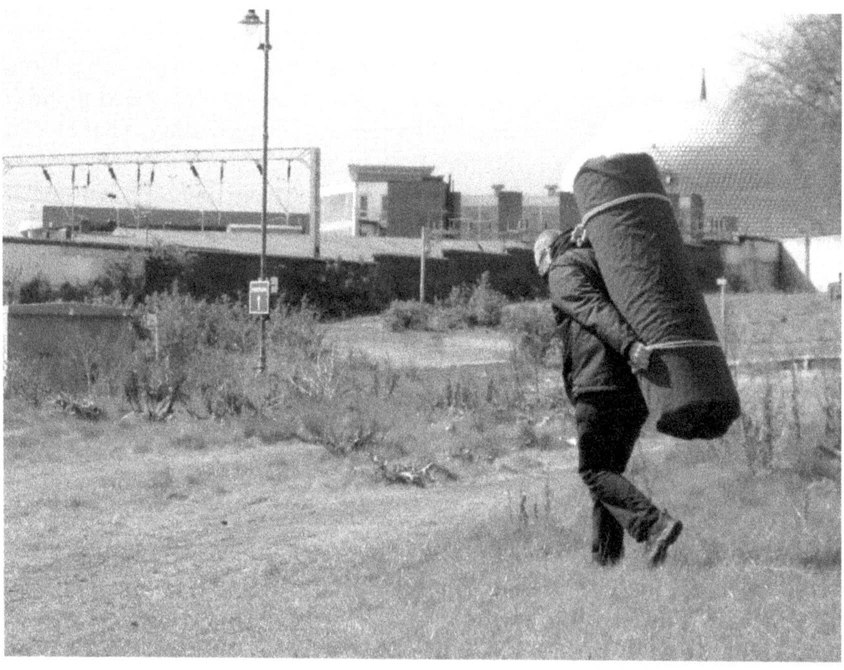

Figure 7.2 Participant carrying a Heavy Object

Source: Nikki Pugh.

made. This turned them into self-playing drums that beat faster the more they were surrounded by the fabric of the city and then slowed as space opened up around them.

Participants were invited to experiment with how and where they carried the Heavy Objects, exploring in what ways these changed how they viewed and interacted with their surroundings.

In the developed Western world of smartphones and other mobile devices, these technologies are often marketed to us on the basis of being increasingly smaller, lighter, and more discreet. The Heavy Objects were in part intended as a challenge to established conventions as to what mobile technologies should be like. In addition to being heavy, they were cumbersome, rigid, and sometimes even slightly abrasive. In use, their physical form continuously draws your attention to your own body, as you make perpetual adaptations in order to control levels of discomfort and maneuverability as experienced from your perspective, as well as the spectacle you provide for those around you.

Thrift (2004) describes an emerging world of ubiquitous background qualitative calculations ('qualculations'), in which increases in computing power—and the tools, systems, and concepts built on it—become paratexts: " 'invisible' forms which structure how we write the world but which generally no longer receive attention because of their utter familiarity" (Thrift, 2004: 585).

By rendering the characteristics of the GPS signal waves in a form that is audible and tangible, the Heavy Objects enable us to perceive an otherwise disregarded invisible data layer that surrounds us. This comes about, however, by also foregrounding our own bodies and the built environment—two things that, despite their constant and physical presence in our everyday lives, also have a tendency to no longer receive attention because of their utter familiarity.

Colony

Possibility Probe (Heavy Object and Built Environment) forms part of an arc of research and making within a larger project known as *Colony*. Here the goal is to make a family of landscape-aware 'creatures' that are carried across the city.

These creatures react to live-processing of GPS signals in much the same way as the Heavy Objects do, but rather than translating multipath error into a drumming noise, they convey it as relative degrees of comfort or distress such that the person carrying them must make navigational decisions in order to keep their creature calm and contented. Thus navigation becomes an exercise much closer in nature to the dérive than to following the predetermined series of lefts, rights, and straight-ons we are becoming acclimatized to through use of sat-navs and route-planners. Within *Colony*, the key questions being constantly evaluated in order to decide which way to go are 'Can you see the sky?' and 'Can you see your friends?'.

The very first iterations of the *Colony* creatures were bundles of bubblewrap with vibration motors along their length where they came into contact with the body of the person carrying them. The consensus from the first groups of people who tried them out was that these were very pleasant because they were light, cuddly, warm and the secretive nature of the communication via the vibrations empowered the carrier to declare the creature's intent without being held to question by the other people accompanying them on the walk.

Before continuing development of the creatures emphasizing these easy wins, it felt important to me to explore what the affordances of a contrary prototype might be—hence the Heavy Objects. I needed to interrogate my assumptions and, since these were so closely linked to properties of mass, texture, and form, I had to do this through a process of making, playtesting, and experiencing.

Learning-Through-Doing and the Enlivening of Public Space

Playtesting is a core aspect of my practice because the tools I build are animated through use—it's hard to understand the possibilities wrapped up within them whilst they lie dormant. Participants gather around an invitation, go out into a public space, and see what happens. Along with my own observations, their responses are fed back into the next iteration of the design process.

Through these participatory phases the development of a project is opened up to external influences, but also made visible to a wider audience of incidental observers who witness the playtesting taking place. Not all of them remain passive observers: during a recent playtesting session in Bristol, strangers approached us asking for photos, to hold the creatures, to 'draw' with the creatures, or simply to ask questions and to chat. The creatures by this stage now had roughly the same shape and dimensions as some of the early bubblewrap bundles, but were made from wood, articulating in response to the GPS data and, whilst not as deliberately uncomfortable or as cumbersome as the Heavy Objects, were designed to be something you had to remain very conscious of in order to carry it (Figure 7.3).

I'm a big fan of urbanist William Whyte's theories of 'triangulation' (an external stimulus facilitating strangers talking together) and of oddball behaviors being an important factor in the enlivening of public social spaces (Whyte, 2009). As we encounter more and more private spaces masquerading as public, and as our inner oddball gets suppressed, it becomes increasingly important to me that apparently silly things are seen to be happening in our shared spaces. For several years now I've been using the word 'malleable' to describe the property of public space that enables me to feel that I can repurpose it to fit me rather than me feeling that I need to conform to it. Similarly, Garrett (2015) describes the "little sprinkling of chaos" he observed at a stadium in Phnom Penh, Cambodia, compared to the lack of diversity in people's engagement with—for example—London's privately owned public spaces.

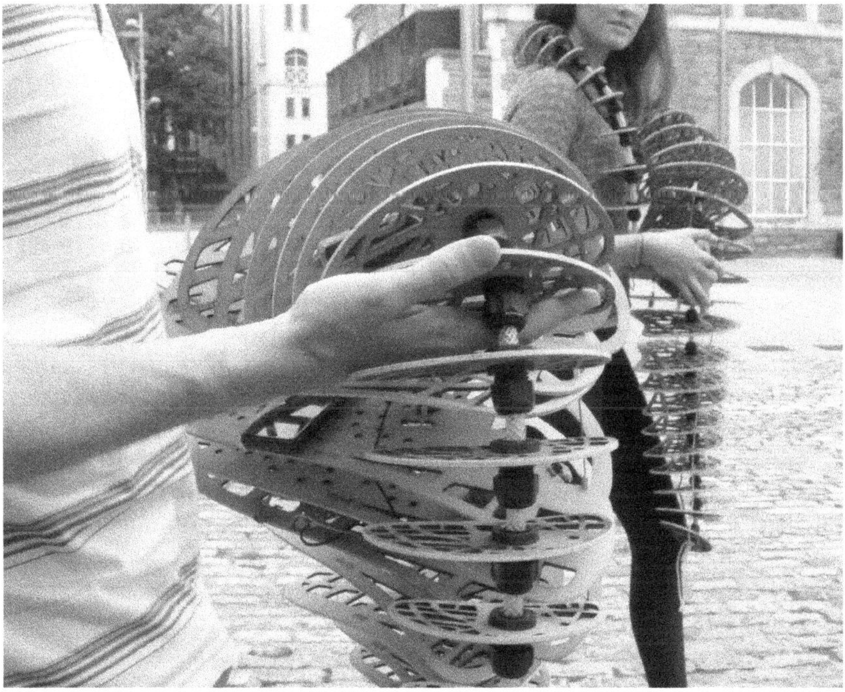

Figure 7.3 Version 9 of the *Colony* creatures, being playtested in Bristol
Source: Nikki Pugh.

Through involvement with pervasive gaming, I observed another, related effect from doing outside-of-the-ordinary things in public places: each time I played a game or engaged with a neglected corner of the city, I was left with a resonance that remained long after the action itself had finished.

I once spent an afternoon with a group of other people playing games and sharing a picnic on a scruffy area of grassland that inhabited the gap between the light industrial factories that had stood there before and the stalled regeneration that had yet to materialize. Several months later I looked out of the window of a passing train and saw construction workers boring into the grassland—*our* grassland—with a massive earth drill. The visceral impact of seeing the site of our playfulness and conviviality being cut into was, I suspect, made even more powerful by the threat of loss—were they finally going to replace this space with office blocks or apartments? I wanted to protect the grassland.

Even when my engagement with a space has been more fleeting, I have observed similar strengthening of my connection to that place. Often this takes the form of a sense of ownership towards it; not in the sense of possession, but a feeling of responsibility, of caretakership. Is this feeling a pre-requisite for more socially engaged actions, and if so, how might this be built upon?

Whilst the ability to even consider the playful co-option of urban space is undoubtedly a privilege, it is not without its frictions. Foremost in my mind as I'm designing are the need for consideration of participants' safety and how events appear from the outside. As game designer Holly Gramazio put it in her 2009 presentation *Exciting Things That Can Go Wrong With Your Pervasive Game*, "your players [running through the street] look like criminals." I am particularly mindful of fears around technology; in particular being able to see wires. It doesn't matter that a large proportion of people are carrying sophisticated electronics and computing in their pockets; simply being able to see wires seems to trigger a combination of memories of movie scenes and fears of terrorist attack. I've yet to encounter this either from the public or security staff whilst out playtesting, but it does come up in conversation and I'm very conscious of it when I've got cables wrapped around me or I'm asking volunteers to do the same. Strategies I use to mitigate the perception of threat include: hiding the wires; use of balloons and silly hats (or their equivalents) as signifiers of play; and informing police or security guards in advance so they know what's happening.

What is it about our routinely carried sleek, black-box versions of this technology that circumnavigates these fears? What exactly is it that gets rendered impotent or invisible by their design or the cultures of use that surround them? Thrift (2004) also reminds us that this is the *modus operandi* of paratexts: Although invisible through their ubiquity, they also work in plain sight to frame and influence how we experience things:

> In other words, number tends to cast the world reciprocally in its image as entities are increasingly made in forms that are countable. Number performs number.
>
> (Thrift, 2004: 590)

How are our mobile technologies casting the world and influencing our expectations and behaviors? How do smartphones perform smartphones, for example? I don't have the answers to these, but they are amongst the questions about our everyday relationships with technology and place that I can pose in a practical, embodied way through my work. Sometimes making people realize there is a question to be asked in the first place is as important as being able to answer it.

Tools for Asking Questions *With*; Spaces for Asking Questions *In*

Handmade mobile technology that strays from convention; sculptural forms that relate to the body whilst extending or hindering its capacity in some way; frameworks for encountering and interacting with public spaces in unusual ways; prototype products that are functional but not yet cast in stone—through the combination of these, a playtesting session is fertile ground for bringing together people with different life experiences and skill sets around an invitation to experiment and make connections.

Galloway's use of 'speculative design ethnography' is described as a way of engaging audiences in debate around important technological, cultural, and socio-political issues (Hendren, 2014), where fiction and artifacts are employed as tools for asking provocative questions. When discussing Design Fiction, Bleeker uses the term 'props' to refer to assemblages that are able to "move the conversation around" on important matters-of-concern (Bleeker, 2009: 85). Rather than objects in isolation, I use multisensual *experiences*—often combining objects, bodies, and spaces—in order to instigate conversations.

My choice to build my own systems is largely born out of the need to make tools that don't already exist: if I want to use them, I have to make them first. Although a significant amount of my comprehension comes about through the making process, it is the playtesting where projects come alive for me. As my practice develops and I gain more and more experience, I am becoming more skilled in shaping these gatherings and crafting the spaces within them for participants to feed into. These spaces are as much about asking new questions as they are about answering existing ones. It is the process of inquiry that feeds me rather than the arrival at a particular solution.

As I type this, plans are afoot for the next iterations of *Colony* (working outside of an institution, the project progresses in phases according to what funding or residencies I can harness). Included in these is the development of a visualization of the journeys participants make whilst they are carrying the creatures. Not only will this show the routes people have taken, but also the changes in the creatures' movements and the expansion and contraction of the group as a whole. The intention is to use a mapping of these data to form an *aide memoir* that participants can gather around and use as a prompt to recall the experience and to share their stories; a debriefing of the event akin to talking about the film you've just watched as you walk home or sit in the pub afterwards. Although I think the walking part of the event would be interesting and entertaining in itself, without that time and space in which to recall, process, and share each individual's experience I feel it would be incomplete and lack the potential to feed forward into something more, for example: to be transposed into larger considerations about the urban built environment and our behavior within it.

Turning our attention to the beginning of the experience, *Colony* hinges on the participants caring enough about the creature they are responsible for that they will alter their navigational priorities in order to keep it in a calm, non-distressed state. This raises a few interesting challenges: how to read what state the creature is in; how to nurture a sense of empathy for the creature; and how to frame the journey with it (and the other creatures) across the city.

Devising alternative interfaces for technology necessarily requires that people then need to learn how to use them. Although this requires an additional investment of time and energy from participants, it is also an opportunity to disrupt expectations and unlearn engrained behaviors. Leading on from this, however, I now face the interesting conundrum that participants across several of my projects have reported they get more from the experience if they are given space in which to figure out the relationships of cause and effect for themselves, rather than having them explained to them. What then is the optimum balance between instruction and self-led discovery? With regards to *Colony*, I believe the issue of empathy is also wrapped up in this process of 'getting to know' the creature.

Framing the journey will be the focus of an upcoming live-lab style residency: Here I will focus on the narrative of the event. For all the talk of digital cities and networked infrastructure, I feel that there is a lot to be gained from superimposing a layer of imagination and storytelling onto our urban landscapes.

The truism 'the pictures are better on radio' can be applied here: A good story can paint the city in Technicolor. Imagination is a high-definition format that rarely suffers from the same glitches and gremlins as other technologies. It also has the benefit of being unique to each participant, leaving space for half a dozen people to have equally rich, vivid experiences that don't have to be exactly the same as each other.

Evolving Methodologies

The knowledge production forming the foundation for my inquiry-led practice has always stemmed from a desire to learn more about the world in which I find myself. Open-ended and growing out from 'what if . . .?' questions, the solitary making phases are complemented and opened up by the more social and public participatory phases of playtesting and discussion that help me gain further insight into what exactly it is that I've made. The making and wearing of something—how it relates to my body and how it affects my experience of my surroundings—is my way in to the investigation, and then input from others helps me to see what the necessary refinements and further possibilities might be that will steer the development of the project into new areas.

As my involvement with academic forms of research increases, I gain new perspectives on my methodologies as well as an increasing desire to reflect more on what is achieved through these exploratory actions. I'm curious to see how I resolve the relationship between traditional research agendas with my existing approach. What will be the tensions between a culture of presenting findings in response to an initial research question and my approach of a research agenda that evolves in response to experimentation and that leaves gaps for people to make their own connections and conclusions?

My (current) ambition for *Colony* is that I will be able to progress it to the point where I can observe numerous groups moving through urban space and analyze their navigational decisions influenced by the interactions of radio waves with the built environment. Regardless of when that stage is reached, the journey of investigation is guaranteed to provide much food for thought along the way.

References

Bleeker, J. (2009). *Design Fiction: A Short Essay on Design, Science, Fact and Fiction*. Available at: http://drbfw5wfjlxon.cloudfront.net/writing/DesignFiction_WebEdition.pdf

Garrett, B. (2015, August 4). The Privatisation of Cities' Public Spaces Is Escalating. It Is Time to Take a Stand. *The Guardian*. Available at: www.theguardian.com/cities/2015/aug/04/pops-privately-owned-public-space-cities-direct-action

Hendren, S. (2014, May 16). *Knitting Bones with Fact and Fiction: A Conversation with Design Culture Lab's Anne Galloway*. [Web log]. Available at: http://abler.kinja.com/knitting-bones-with-fact-and-fiction-a-conversation-wi-1551078211?rev=1400196889

Thrift, N. (2004). Movement-space: The Changing Domain of Thinking Resulting from the Development of New Kinds of Spatial Awareness. *Economy and Society*, 33(4), 582–604. doi: 10.1080/0308514042000285305

Whyte, W. H. (2009). *City: Rediscovering the Center*. Philadelphia: University of Pennsylvania Press. Available at: https://books.google.com/books?id=OH6y2QdcUqkC&pgis=1

8 (Re)Envisioning the Anti-Urban

Artistic Responses in the Walking With Wordsworth and Bashō Exhibition

Mike Collier and Kevin Hannam

In this chapter, we discuss the exhibition *Wordsworth and Bashō: Walking Poets*, held at Dove Cottage in the Lake District, UK, in 2014, in the wider context of Romantic anti-urbanism. The main aim of the exhibition was to compare the work of two famous poets in the context of the unique landscape of the English Lake District. In the text that follows, we show how these poets are linked through the agency of walking with the wider context of anti-urbanism and then discuss contemporary artistic responses to the poets. We firstly discuss the relevance of this chapter to the book as a whole and then outline the cultural importance of walking in both the UK and Japan. Next we introduce the walking methods of both of the poets, Wordsworth and Bashō, demonstrating that their work was formally and emotionally structured by their practices of walking. We then discuss the ways in which walking informed the making of the artistic works in the exhibition itself and we conclude by reflecting further on walking and anti-urbanism more generally.

Since the Romantic Movement, cities have been re-envisioned through anti-urban sentiments and these have been expressed politically by different classes of people in many Western and non-Western countries. However, such sentiments have been harnessed, at various times and places, by right-wing political agendas to put forward a particular sense of community and a vision of rural idealism that is seen as being somehow 'lost' through urbanization, suburbanization, and the rise of 'ghettoes' (Krim, 1992). In the Belgian context, De Decker, Kesteloot, De Maesschalck, and Vranken (2005: 157) have traced the current wave of anti-urbanism expressed through the rise of right-wing politics back to the "geography of the nineteenth-century industrial revolution and its social consequences." Similarly, Marchand and Cavin (2007) have examined the origins of anti-urban politics from the eighteenth-century social commentaries of Rousseau through to right-wing planning perspectives developed during the Second World War in France and Switzerland. More recently Cavin (2013) has suggested that this anti-urban bias still resides in Switzerland in residents' attitudes towards cities. Such examples highlight that anti-urban politics remain current in the political agenda in many countries.

Nevertheless such sentiments have been critiqued by both academics and artists who seek to re-envision the anti-urban as a way to understand the relations between society and nature. In particular, studies of the politics of walking has enabled artistic interventions within the human landscape (Gros, 2014). One such critique has been through the development of artistic interventions that highlight ways of walking in nature in order to re-evaluate anti-urban sentiments as a different way of looking at the relations between urban and rural landscapes rather than as being shoe-horned into the service of right-wing political agendas.

The Politics of Walking in the UK and Japan

Walking has been a central part of historical and contemporary artistic responses to and interventions within the human landscape since at least the eighteenth century (Gros, 2014). More recently, walking has also become an integral part of contemporary artistic methods and politicalized responses to changes in landscapes and how they are used. For example, Myers (2011: 183) has highlighted how narrative walking practices, "or modes of conversational activity set in motion by the conditions of wayfinding," offer a means of intervening in the politics of mobility. She discusses how the artwork "way from home" provided an "interventional methodology for eliciting and representing the transnational experiences, affects and significances of place for refugees and asylum seekers across the UK" (Myers, 2011: 183).

Walking is thus also a medium through which we can 'expressively' take direct action; walking as performative agency. In the UK, such 'action' has manifested itself across the twentieth century including the Jarrow March in 1936 (a walk that was undertaken again 60 years later by the artist Tim Brennan). The 'walk' was a means of direct action in the political, social, and geographical landscape (Collier, 2014). The relationships between art, walking, and different urban and non-urban environments are thus complex in terms of their poetics and politics (Pinder, 2011).

The theme of anti-urbanism was a core feature of nineteenth-century Romanticism which was central to the 'framing' of nature as a spiritual realm (Blanning, 2010). The Romantic Gaze constructed through Romantic artists' portrayals of nature as a realm of the solitary and the spiritual became challenged by the presence of other leisure and tourism walkers such as the ramblers described above (Macnaghten & Urry, 1998; Whiting & Hannam, 2014). However, the natural realm so central to the Romantic ethic and constructed in Romantic ideology as an authentic arena, separate and distinct from the emerging modern urban world, itself has ironically become an object of consumption, and a space of occupation, by the embodiment of modern alienation, namely the tourist (Buzard, 1993; Stokes, 2012; Whiting & Hannam, 2014).

Furthermore, the irony of this was that the practices and artistic productions of many Romantic writers and poets themselves became key inscriptions of new meanings of 'sublime' landscape and were distilled into the guide books of the emergent tourism industry (Urry, 2002). In this sense we can see the emergence of one particular strand of Romantic practice, that of travel in sublime nature, being appropriated and transformed through cultural emulation and nascent consumer capitalism into a popular activity in the form of nature-based tourism (Whiting & Hannam, 2014).

In this context, the attraction of the Lake District landscape is complex. Experiences are, in part, the "product of a revolution in cultural values during (mainly) the second half of the eighteenth century, which transformed perceptions of the aesthetics of coastal as well as mountain scenery" (Walton, 2013: 33). Prior to this time, most people had viewed the mountains as grim, frightening, and inhospitable—places to avoid (Nicholson, 1997). Better transport links as well as changes in the cultural appreciation of areas such as the Lake District led to an increasing number of tourists being drawn to the Cumbrian Fells. The Romantic poets, especially Wordsworth who was born in Cockermouth and lived and worked in the Lake District for much of his life, did much to popularize not just an aesthetic, but also an immersive appreciation of these hills, valleys, and lakes. Indeed, Hanley (2013: 113) calls the Lake District 'The Region of Wordsworth's song' and explains that that the Lake District has become known as 'Wordsworthshire' because:

> It is more natural and legitimate to associate Wordsworth with certain parts of England than any other great writer. And for three reasons: He spent the greater part of his life in one district; he drew much of his scenery and human character from the district and used its place-names freely in his poems; and both he and his sister left considerable records of his time and places of composition.

Furthermore, Wordsworth took part in campaigns to 'conserve' the Lake District "as places or enclaves distant and protected from science, industry and the operation of power" (MacNaghten & Urry, 1998: 24). Thus Wordsworth was central to the envisioning of an anti-urban Romantic discourse (Luckin, 2006). However, there is also considerable irony in the fact that it was the rise of the railway and related technology, which drove the Industrial Revolution (and gave rise to the slums), that also enabled people from the city to travel to the countryside in ever increasing numbers, to free themselves from the "impact modern life was having on the body," a body that was being increasingly and physiologically "undermined, superseded and inscribed by technology" (Lewis, 2001: 64).

Wordsworth also encouraged 'meanders' from the everyday routes. In both his *Guide to the Lakes* and in his poetry, Wordsworth encouraged visitors to stray from the beaten track and experience the Lake District's hidden natural treasures, its waterfalls, crags, and fell tops, the changing seasons, and the working life of the people who lived there, rather than a set of viewing 'stations' on a Picturesque Tour. Wordsworth argued that the Lake District demanded a different eye, one which is not threatened or frightened by the relatively wild and untamed nature. It requires "a slow and gradual process of culture" (Wordsworth, 1884: 193, cited in MacNaghten & Urry, 1998: 211).

The distances walked by the Romantic writers of the period were considerable: William Hazlitt claimed to walk 40 or 50 miles a day, De Quincey walked 70 to 100 miles a week, and Keats covered 642 miles during his 1818 tour of the Lakes and Scotland (Wallace, 1993: 166–167). By the middle of the nineteenth century "the very highest echelon of English society regarded pedestrian touring as a valuable educational experience" (Wallace, 1993: 168). Walking had become "particularly associated with 'the intellectual classes' who had begun to develop quite complex justifications, [such as] a 'peripatetic theory' based upon the way that the pedestrian is supposedly re-created with nature" (MacNaghten & Urry, 1998: 355).

Since the Romantic period, walking as a leisure activity emphasizing this sense of freedom has continued to grow in popularity in the UK (Edensor, 2000). The idea that the countryside is the preserve of the elite was famously contested by the largely working-class national Federation of Rambling Clubs which formed in 1905 and emphasized collective walking and contesting the ownership of land. Rambling clubs soon sprang up in the north and began campaigning for the legal 'right to roam' as much of the countryside was privately owned and trespass illegal. Access improved with the introduction of the National Parks and Access to the Countryside Act in 1949, and in 1951 the first national park in the UK was created. More national parks followed, helping to improve access for many outdoors enthusiasts. The establishment of The Countryside and Rights of Way Act 2000 further extended the right to roam in England and Wales. Since the turn of the century, there has been a growth in the number of 'Walking Festivals' in the UK where walking can help define a sense of place (touristically). Walking is understood to produce intense feelings of liberation or refreshment, a stronger connection with what is elemental, and a slowing in the otherwise hectic pace of life (Lorimer & Lund, 2008).

However, in spite of the continued interest in walking in the UK, it is apparent that people from different backgrounds may also appreciate walking in the Lake District differently. It is mainly an Anglophone, even an English, attraction. The eighteenth-century transformation of attitudes to landscape and scenery in Western Europe was absent in many other societies who viewed the experience of this kind of landscape as a 'native English', or at least Anglophone, cultural possession which was not expected to be shared but which has nonetheless been contested by different migrant groups through artistic engagement (see Tolia-Kelly, 2008; Myers, 2011).

Although the early twenty-first century witnessed a boom in Japanese visitors to the Lake District, it was a boom focused not on an appreciation of Wordsworth's poetry or the scenery it describes, but rather on the whimsical delights of Beatrix Potter's Peter Rabbit books, which were translated into Japanese in 1971 and had become much-loved school texts for the learning of basic English (in Japan) (see Squires, 1993; Iwashita, 2006). Whilst many Anglophile visitors to the Lake District are keen to experience it by walking through and in it, encouraged by a series of highly informative and detailed guide books, as well as the cultural associations brought forth by Wordsworth and the Romantic Poets, few Japanese tourists come to the Lake District to walk in the landscape, even though there is a long-standing culture of walking in Japan, particularly in mountains (Fukada, 2015). However, it seems that this type of appreciation was/is not being promoted abroad. Walton (2013) points out that "Cumbria Tourism and its consultants tend to take landscape aesthetics and literary associations as 'read', and seek to further expansion of tourism by other routes especially the 'stately homes of Levens, Holker and Lowther Castle." The Head of Marketing & Communications for the Wordsworth Trust, Paul Kleian, further explained that, of course, whilst the landscape itself has little direct finance, the buildings (owned by the National Trust and other private organizations) do have promotional budgets. However, Kleian argued that The National Park, Visit Britain, and Cumbria Tourism were becoming increasingly aware (over the last two years especially) that for visitors from large urban conurbations like Tokyo, fresh air and being away from urban areas was a major selling point, and they welcomed the chance to explore a cultural landscape without having to wear a pollution mask.

Walking Poets

The most obvious 'similarity' between the Wordsworths and Bashō is that they were inveterate walker-poets. Wordsworth's contemporary Thomas De Quincey reckoned that Wordsworth walked a distance of 175,000 to 180,000 miles in his lifetime and it could be argued that walking creates one of the main themes around which *The Prelude* is constructed. Two of the most important experiences related in *The Prelude*, for example, found their origins in mountain walks (for instance when overawed by the Ravine of Gondo, the "Gloomy Pass," or during a night walk to the top of Snowdon). Walking was important to Wordsworth because it created an *interaction* between the traveler and the landscape (McCracken, 1988). It was similarly important to Dorothy Wordsworth, who walked incessantly throughout much of her life, both on her own and with companions, recording her observations and encounters in the pages of her *Grasmere and Alfoxden Journals*.

The same rationale for walking and writing could equally apply to Bashō, who made not one but several journeys in Japan. "He sought to experience first-hand beautiful scenes such as Mount Yoshino, Sarashini, and the pine-clad islands of Matsushima" (Barnhill, 2005: 5). His first journey, in 1684, was described in *Nozarashi Kikō* ('A Weather Beaten Journey').

Other journeys and journals followed (including, in 1687, *Oi no Koumin*, 'The Records of a Travel-Worn Satchel' and, in 1688, *Sarashina Kikō*, 'A Visit to Sarashina Village'). His art reached its greatest form in 1689 in his masterpiece *Oku no Hosomichi*, 'The Narrow Road to the Deep North.' In this poem/travel book, he recounts his last long walk, completed with his disciple Sora, some 1,200 miles covered over five months beginning in May 1689. The Japanese term *oku* refers to the northern backcountry of the main Japanese island of Honshu, and it also means 'deep' in the sense of interior, such as the depths of a mountain and spiritual depths (Barnhill, 2005).

Although both the Wordsworths and Bashō have been labeled 'nature poets,' this could be, perhaps, a little misleading, because they were also very much concerned with people or some form of 'cultured nature' (Keene, 1982). In Bashō's prose, for instance, we encounter a wide variety of people he met 'on the road'—each different and individual. A glance at the subjects of some of Wordsworth's poems ('The Sailor's Mother'; 'Beggars'; 'The Discharged Soldier,' and 'The Leech Gatherer') reveal his interest, too, in the people of the road he met when walking. Their poetry rejects the simple definitions of the landscape as an "alien and formless substrate awaiting the imposition of human order" (Ingold, 2000: 191), or of "landscape as a 'cultural image', that is as first and foremost a symbolic representation . . . a '*way of seeing*' that assumes and reproduces a fundamental distinction between the *ideas of culture* and the *matter of nature*" (Wylie, 2007: 154). Rather, in their poetry, subject and object, mind and body, and, especially, culture and nature, are conjoined rather than made distinct.

Artistic Responses in the Walking With Wordsworth and Bashō Exhibition

Elder (2014: 28) suggests that although "there are dramatic gaps between the poetic worlds of Wordsworth and Bashō," it is all the more impressive that there are

> deep resonances between these poets that have contributed to making Bashō such an important influence on western poets today . . . and . . . Wordsworth so beloved in Japan. . . . They are kindred writers in part through their shared calling to the margins of their societies, the shifting edges between nature and culture where insight may germinate. Bashō's narrow path to the interior, *Oku no Hosomichi*, anticipates Wordsworth's choice, at the beginning of 'Michael' and elsewhere, to depart 'from the public way'.

Encouraged by this synergy, focused around walking, between two writers from different cultures who had never visited the other's country, colleagues in the WALK research center at the University of Sunderland[1] and

Jeff Cowton, Curator of the Wordsworth Trust, decided to embark on a collaboration, invited contemporary artists to respond to the manuscripts of the Wordsworths and Bashō—to what Cowton called "the texture of thought" that the manuscripts revealed. We aimed to examine these contemporary responses within the comparative context of Western and Eastern attitudes to the natural world and walking, and to explore the nature of collaboration itself. For although many will be aware that Bashō often worked and wrote collaboratively, undertaking his journeys with companions, it may come as a surprise to realize that Wordsworth was also a creative collaborator (with Coleridge, his sister Dorothy, and his wife, Mary) rather than the solitary poet. It was in this spirit of collaboration that artists were invited to work collaboratively—and in the spirit of cultural exchange to encourage artists from the UK and Japan to work together.

Scottish artists Alec Finlay and Ken Cockburn had already had an interest in Bashō and haiku poetry. Both had previously traveled on 'The Road North' through their Scottish homeland, guided on their journey by Bashō's prose poem *Oku no Hosomichi* ('Narrow Road to the Deep North'). Following Bashō and Sora, their journey took in 53 'stations,' from Pilrig to Pollokshields via Berneray, Glen Lyon, Achnabreck, and Kirkmaiden. They left Edinburgh on 16 May 2010, the same date that Bashō and Sora departed Edo in 1689, and finished their journey at Glasgow's Hidden Gardens in May 2011 (Finlay & Cockburn, 2014).[2] This journey also provided also them with ideas for their work in the exhibition (Figure 8.1). Finlay's

> simple 'word- mntn' are constructed from wood cubes that, on one level, we may associate with children's building-blocks. Deceptively playful, the 'word-mntn' in fact mask complex layers of literary, artistic, personal and philosophical association. Like much of Finlay's work, they are influenced by Bashō's pared back haiku and exemplify the artist's typical concerns with human engagement in landscape.
>
> (Collier, 2014: 86)

In *The Mulberry Coat*, Autumn Richardson draws upon the natural landscape of Japan (Figure 8.2).

> Before setting off on the journeys from which his poetry sprang, Bashō required hand-made coats, hats and shoes, all of which were made from the very plant life that inspired his writing. . . . 'The Mulberry Coat' is an artistic re-imagining of the garment that Bashō might have worn, made from untreated, unsized kozo (mulberry) paper (made by paper-maker Nao Sakamoto), and sewn with cotton thread. By focusing on the paper clothing itself, Richardson emphasises the materiality of Bashō's writing process, and the value of paper as a form of protection, a means of recording his thoughts, and a direct form of contact with the landscape itself.
>
> (Richardson, 2014: 124)

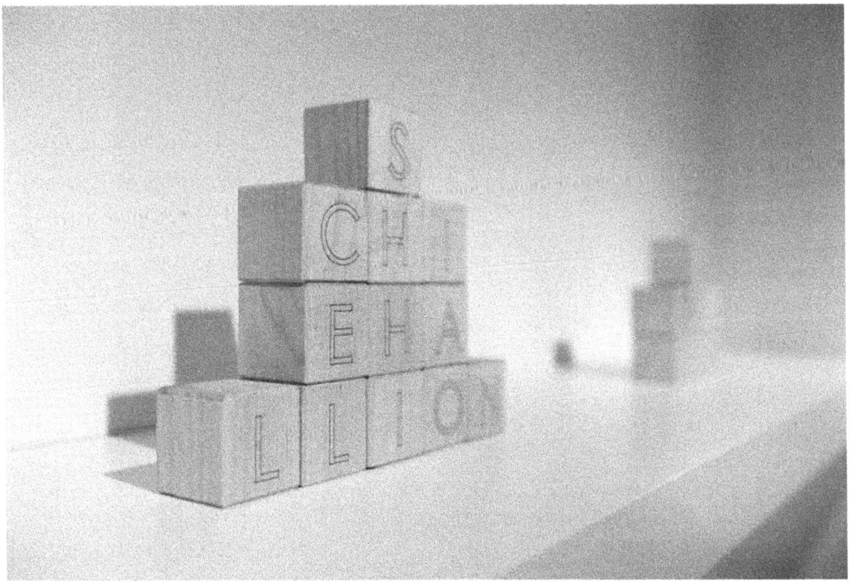

Figure 8.1 Alec Finlay, *word-mntn* (Schiehallion), 2014, maple cubes, each 4 x 4 cm
Source: Photo by Colin Davison—Rosella Studios.

In Andrew Richardson's work *The Ghostly Language of Ancient Earth*, extracts from Wordsworth's *The Prelude* are mapped onto a three-dimensional interactive map of the landscape of Grasmere and the surrounding area (see Figure 8.3). The interactive screen version of the map allows viewers to explore the text-landscape, to view it from different angles and perspectives, or even to follow, to walk along, a specific 'path' of words. These word paths are directly linked to the shape of Wordsworth's poetry, as the number and length of words in each verse line determines the path's change of direction. In re-presenting the contours of the landscape Wordsworth wrote about, Richardson returns the poet's words literally and metaphorically to the topography that was so much a part of the poet's life and work. In turn, Richardson allows us to explore the contours of this landscape through touch-screen. We are able to track our own path through the topographical lines of the map, whilst simultaneously following the poetic lines of Wordsworth's most famous literary masterpiece. In so doing, our fingers take us for a walk both literally and imaginatively.

Work by Mike Collier relates directly to walks undertaken 'in the footsteps' of others, including Bashō and Wordsworth, in Kurabane and Nikko in Japan following short sections of Bashō's route in the 'Narrow Road to the Deep North' as well as many of the same Lake District routes

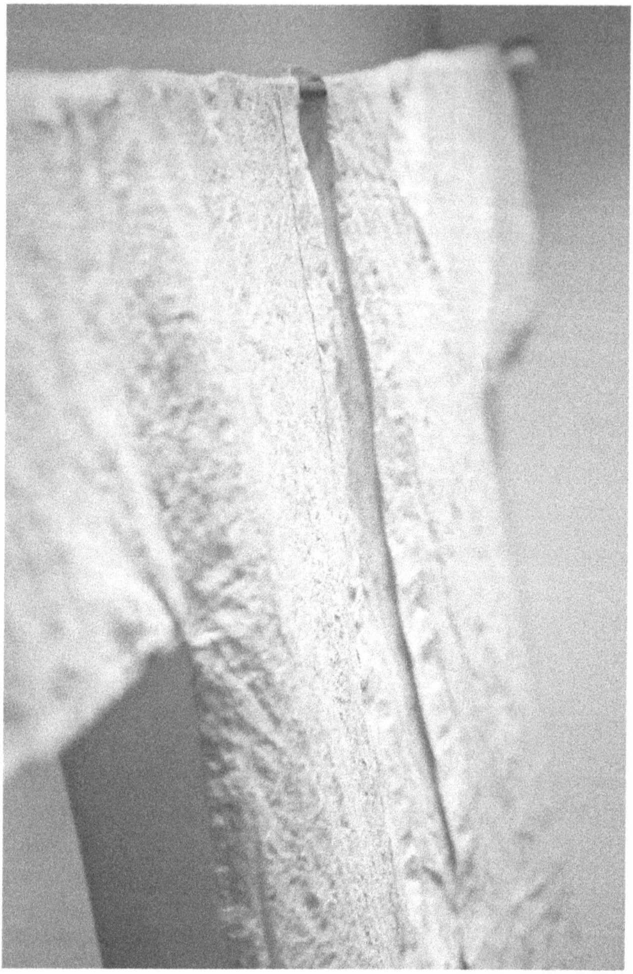

Figure 8.2 Autumn Richardson, *The Mulberry Coat*, 2014, sewn kōzo paper, 123 cm x 128 cm

Source: Photo by Colin Davison—Rosella Studios.

described by Dorothy Wordsworth in her journals (see Figure 8.4). Writing in the catalogue for the exhibition, Carol McKay (2014: 75) explains that in his

> creative reworking, Collier has deliberately selected journal entries that describe favourite routes Dorothy followed—and ones he has walked a number of times: an ascent of Fairfield foiled by weather, followed two days later by a walk 'upon Helvellyn, glorious, glorious sights'.

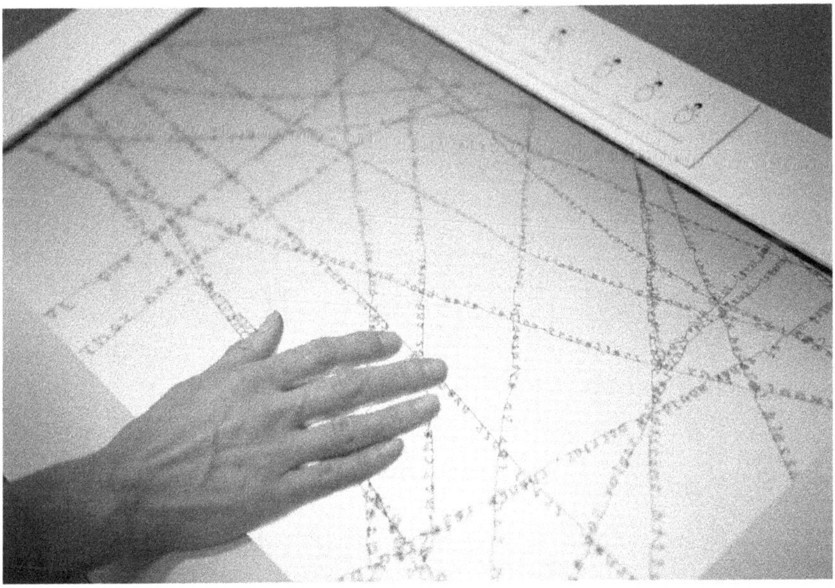

Figure 8.3 Andrew Richardson, *The Ghostly Language of Ancient Earth*, 2014, interactive map and digital prints, 1280 x 720 screen pixels on a 23-inch monitor. Collier, M. (Ed.) (2014). *Wordsworth and Bashō: Walking Poets*. Sunderland and Grasmere: AEN and the Wordsworth Trust

Source: Photo by Colin Davison—Rosella Studios.

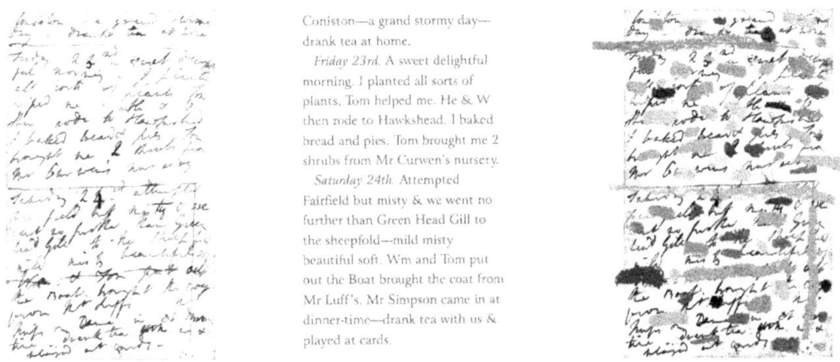

Figure 8.4 Mike Collier, *The Texture of Thought: Fairfield (From Dorothy Wordsworth's Journal Entry, 23rd & 24th October 1801)*, unison pastel onto digital print, 2014, 35 x 65 cm

Walking is also integral to Brian Thompson's artistic practice. He is continuously fascinated by the physical journeys we make in and through places, and how these are mapped, recorded, and valued, and he often walks with friends whose initials are noted in each work's title. Sensitivity to the

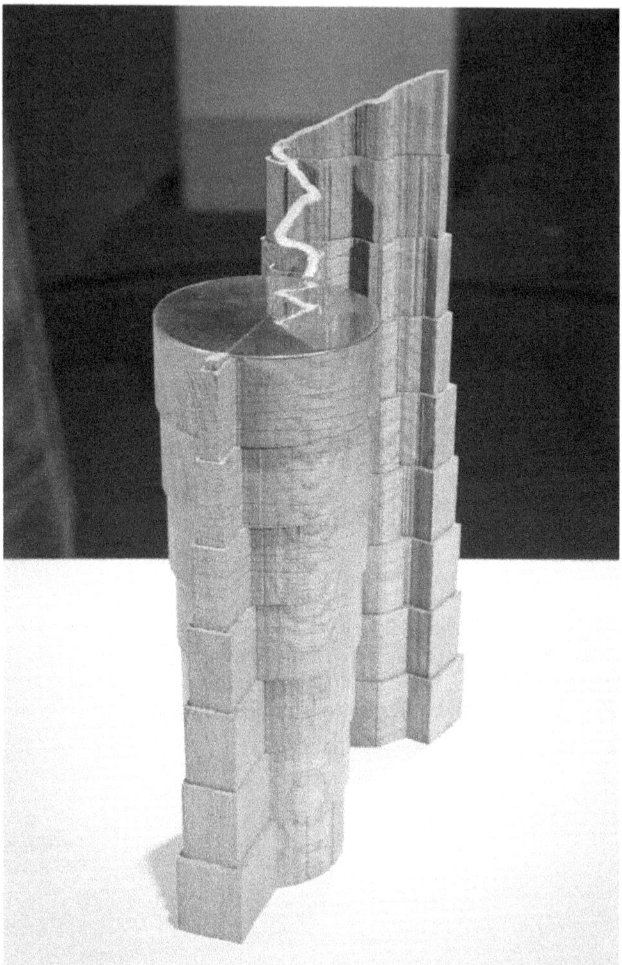

Figure 8.5 Brian Thompson, *Over the Ferry to the Station (with J, E & B)*, 2014, oak and copper, 29 x 17 x 8 cm

Source: Colin Davison—Rosella Studios.

physicality and aesthetics of materials is also central to his work. Each individual sculpture is fabricated from materials that have some relevance to the particular walks that inspired them. Three of the smaller sculptures exhibited here were inspired by walks in and around the Lake District made 'in the footsteps of Wordsworth,' whilst two more had their origins in walks made in Japan with Mike Collier whilst following paths once trodden by Bashō (Figure 8.5).

Conclusion

Wordsworth and Bashō: Walking Poets opened in May 2014 at Dove Cottage. It re-explored traditions of 'nature' poetry through both a Western and a Japanese cultural context. Manuscripts and facsimiles were reinterpreted by contemporary artists working in a wide range of media, including fine art, music, textiles, and sculpture. The exhibition ran for six months, closing in November 2014, after 33,000 visitors. The exhibition demonstrated the complex interactions between poetic interpretations of nature and responses to the modern city through walking, and contemporary artistic responses to the poetry, walking in the Lake District, and more generally the re-envisioning of the anti-urban. While anti-urban sentiments are frequently drawn into the right-wing political arena, this exhibition suggests that the rural idealism that underpins such sentiments can be challenged by artistic interventions that lead to a re-envisioning of urban mobilities.

Notes

1. WALK (Walking, Art, Landskip and Knowledge) is an interdisciplinary research center at the University of Sunderland looking at how cultural practitioners creatively engage with the world as they walk through it.
2. See www.theroadnorth.co.uk/.

References

Barnhill, D. (2005). *Bashō's Journey*. Albany, NY: SUNY Press.

Blanning, T. (2010). *The Romantic Revolution*. London: Weidenfield & Nicholson.

Buzard, J. (1993). *The Beaten Track: European Tourism, Literature and the Ways to 'Culture' 1800–1918*. Oxford: Oxford University Press.

Cavin, J. S. (2013). Beyond Prejudice: Conservation in the City: A Case Study from Switzerland. *Biological Conservation*, 166, 84–89.

Collier, M. (Ed.) (2014). *Wordsworth and Bashō: Walking Poets*. Sunderland and Grasmere: AEN and the Wordsworth Trust.

De Decker, P., Kesteloot, C., De Maesschalck, F., and Vranken, J. (2005). Revitalizing the City in an Anti-Urban Context: Extreme Right and the Rise of Urban Policies in Flanders, Belgium. *International Journal of Urban and Regional Research*, 29(1), 152–171.

Edensor, T. (2000). Walking in the British Countryside: Reflexivity, Embodied Practices and Ways to Escape. *Body and Society*, 6(3–4), 81–106.

Elder, J. (2014). Borderers. In M. Collier (Ed.), *Wordsworth and Bashō: Walking Poets* (pp. 28–30). Sunderland and Grasmere: AEN and the Wordsworth Trust.

Finlay, A., and Cockburn, K. (2014). *The Road North*. Bristol: Shearsman.

Fukada, K. (2015). *One Hundred Mountains of Japan*. Honolulu: University of Hawaii Press.

Gros, F. (2014). *A Philosophy of Walking*. London: Verso.

Hanley, K. (2013). The Imaginative Visitor: Wordsworth and the Romantic Construction of Literary Tourism in the Lake District. In J. K. Walton and J. Wood

(Eds.), *The Making of a Cultural Landscape: The English Lake District as Tourist Destination, 1750–2010* (pp. 113–132). Farnham: Ashgate.

Ingold, T. (2000). *The Perception of Environment: Essays on Livelihood, Dwelling and Skill.* London: Routledge.

Iwashita, C. (2006). Media Representation of the UK as a Destination for Japanese Tourists. *Tourist Studies*, 6(1), 59–77.

Keene, D. (1982). *Appreciations of Japanese Culture.* Tokyo: Kodansha International.

Krim, A. (1992). Los Angeles and the Anti-tradition of the Suburban City. *Journal of Historical Geography*, 18(1), 121–138.

Lewis, N. (2001). The Climbing Body. In P. MacNaghten and J. Urry (Eds.), *Bodies of Nature* (pp. 58–77). London: Sage.

Lorimer, H., and Lund, K. (2008). A Collectable Topography: Walking, Remembering and Recording Mountains. In T. Ingold and J. L. Vergunst (Eds.), *Ways of Walking: Ethnography and Practice on Foot* (pp. 318–345). Aldershot: Ashgate.

Luckin, B. (2006). Revisiting the Idea of Degeneration in Urban Britain, 1830–1900. *Urban History*, 33(2), 234–252.

MacNaghten, P., and Urry, J. (1998). *Contested Natures.* London: Routledge.

Marchand, B., and Cavin, J. S. (2007). Anti-urban Ideologies and Planning in France and Switzerland: Jean-François Gravier and Armin Meili. *Planning Perspectives*, 22(1), 29–53.

McCracken, D. (1988). *Wordsworth and the Lake District: A Guide to the Poems and Their Places.* Oxford: Oxford University Press.

McKay, C. (2012). *Walking Otherwise: One Foot After Another.* Paper presented to the Association of Art Historian's (AAH) Conference, Milton Keynes, 29–31 March. Copy available from the author.

Myers, M. (2011). Walking Again Lively: Towards an Ambulant and Conversive Methodology of Performance and Research. *Mobilities*, 6(2), 183–201.

Nicholson, M. H. (1997). *Mountain Gloom and Mountain Glory.* Seattle: University of Washington Press.

Pinder, D. (2011). Errant Paths: The Poetics and Politics of Walking. *Environment and Planning D: Society and Space*, 29, 672–692.

Richardson, A. (2014). The Mulberry Coat. In Collier, M. (Ed.), *Wordsworth and Bashō: Walking Poets* (pp. 124–127). Sunderland and Grasmere: AEN and the Wordsworth Trust.

Squires, S. (1993). The Cultural Values of Literary Tourism. *Annals of Tourism Research*, 21, 103–120.

Stokes, E. (2012). Sign, Sensation and the Body in Wordsworth's 'Residence in London'. *European Romantic Review*, 23(2), 203–223.

Tolia-Kelly, D. P. (2008). Motion/Emotion: Picturing Translocal Landscapes in the Nurturing Ecologies Research Project. *Mobilities*, 3(1), 117–140.

Urry, J. (2002). *The Tourist Gaze.* London: Sage.

Wallace, A. (1993). *Walking, Literature and English Culture.* Oxford: Clarendon Press.

Walton, J. K. (2013). Setting the Scene. In J. K. Walton and J. Wood (Eds.), *The Making of a Cultural Landscape: The English Lake District as Tourist Destination, 1750–2010* (pp. 31–48). Farnham: Ashgate.

Whiting, J., and Hannam, K. (2014). Journeys of Inspiration: Working Artists' Reflections on Tourism. *Annals of Tourism Research*, 49(1), 65–75.

Wordsworth, W. (1884). *Illustrated Guide to the Lakes.* Exeter: Webb & Bower.

Wylie, J. (2007). *Landscape.* London: Routledge.

Chapter 9, Antonia Hernández

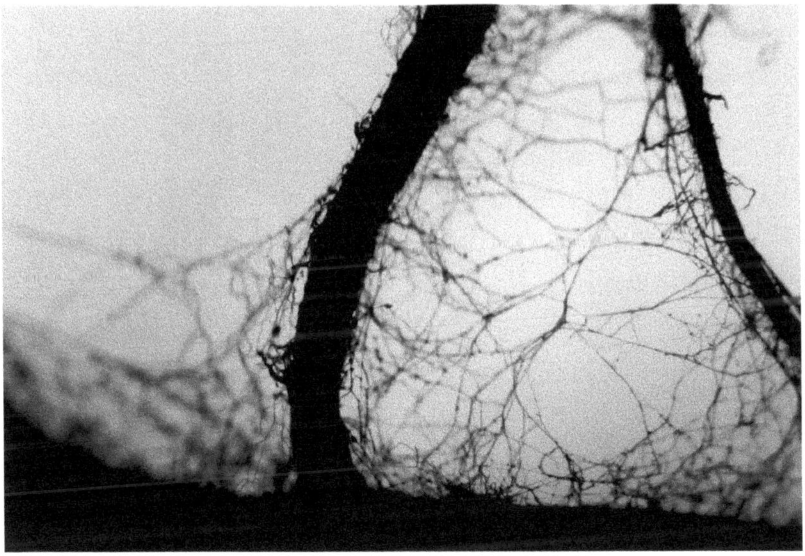

Figure 9.6 The Moldy Strategy. Domestic entity captured with a digital microscope

Source: Antonia Hernández.

Chapter 11, Ulrike Boskamp and Annette Kranen

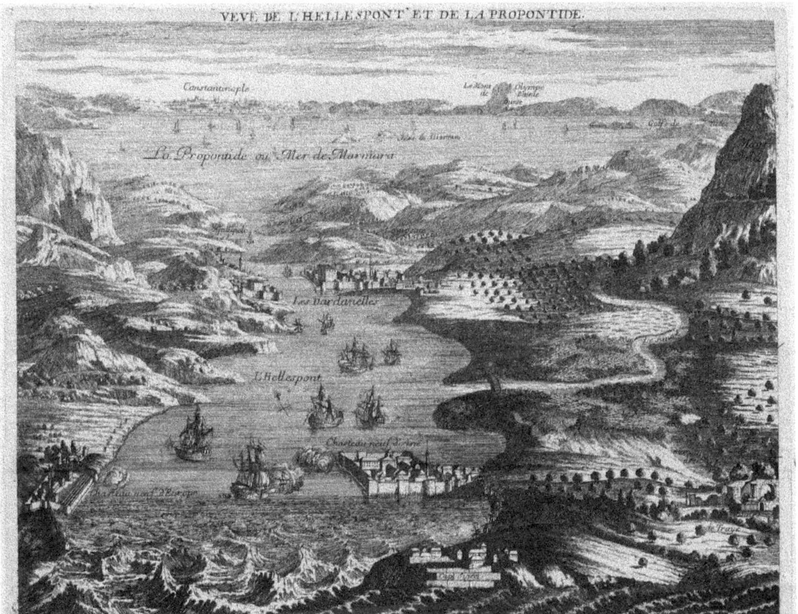

Figure 11.1 Guillaume Joseph Grélot, *Veue de l'Hellespont et de la Propontide*, from Grélot, *Relation nouvelle d'un voyage de Constantinople*, Paris 1680, 41

Source: Hellenic Library, Alexander S. Onassis Public Benefit Foundation, Photo © Aikaterini Laskaridis Foundation.

Chapter 6, David Pinder

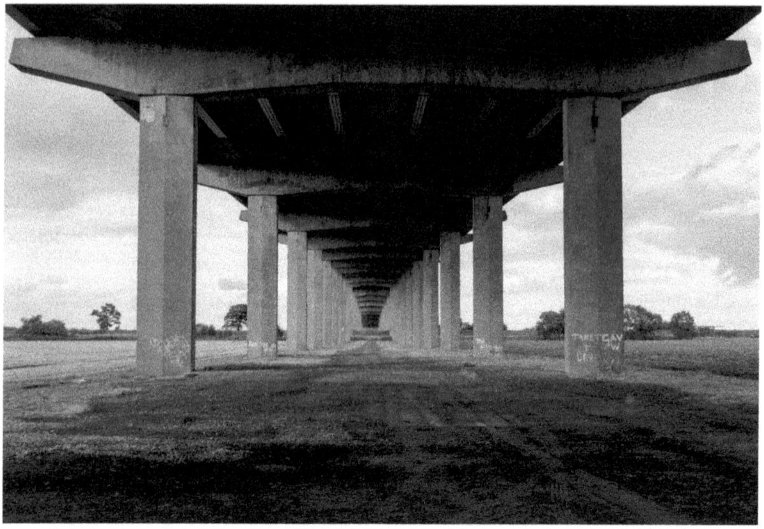

Figure 6.1 "Spectral sounds" under the M62 Ouse bridge at Goole, East Yorkshire, UK: "a few steps into the bridge's shadow remove all vehicular movements from view, and in the sudden stillness the senses open to the booming, cracking, clanking, noises above" (Davies, 2012: 20)

Source: © Phil Hunter/Alamy Photo.

Chapter 7, Nikki Pugh

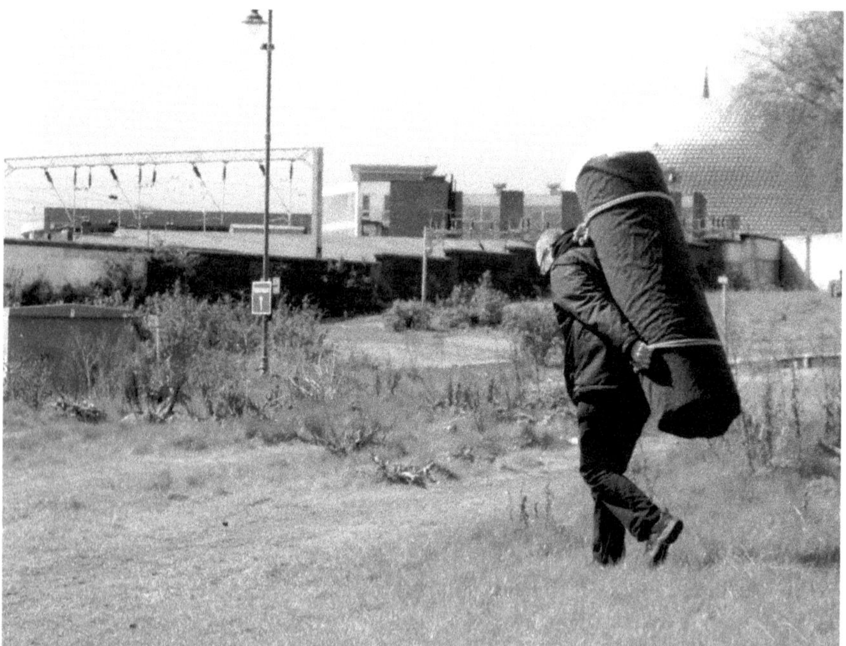

Figure 7.2 Participant carrying a Heavy Object

Source: Nikki Pugh.

Chapter 8, Mike Collier and Kevin Hannam

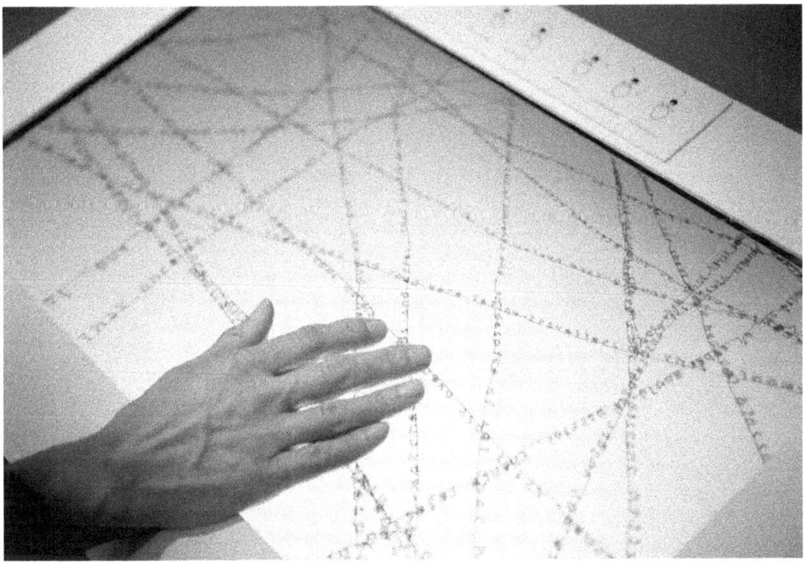

Figure 8.3 Andrew Richardson, *The Ghostly Language of Ancient Earth*, 2014, interactive map and digital prints, 1280 x 720 screen pixels on a 23-inch monitor. Collier, M. (Ed.) (2014). *Wordsworth and Bashō: Walking Poets*. Sunderland and Grasmere: AEN and the Wordsworth Trust

Source: Photo by Colin Davison—Rosella Studios.

Chapter 12, Lee Lee

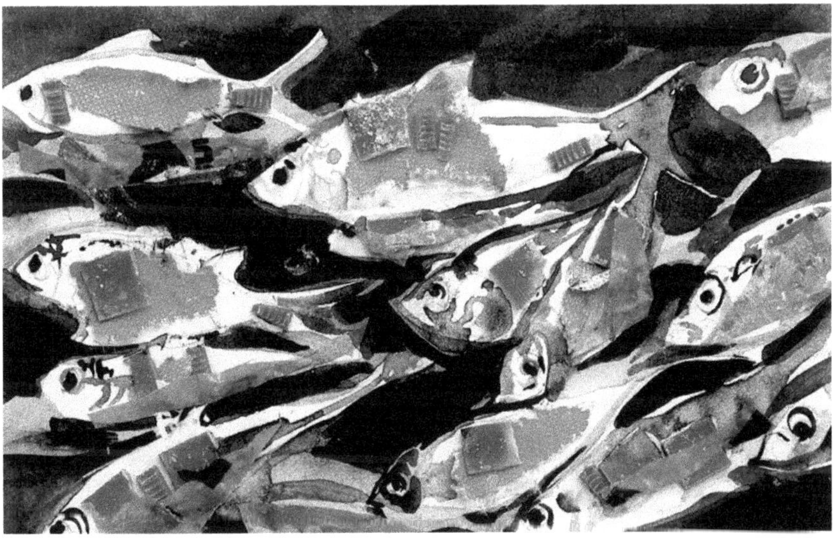

Figure 12.1 The Debris Project, Cosmobilities Installation, Aalborg University. This image of alewives, painted by Lee Lee, represented the Debris Project in an exhibition on Biodiversity and Extinction hosted by Art Science Collaborations at the New York Hall of Science

Source: Lee Lee.

Chapter 4, Peter Peters

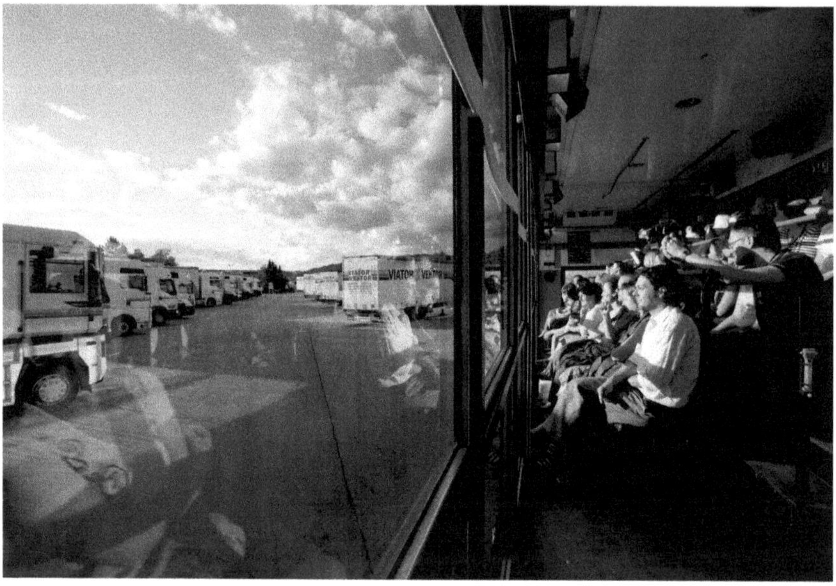

Figure 4.1 *Cargo Sofia-X* by Rimini Protokoll (Stefan Kaegi)

Source: Photo © Nada Žgank at International festival Mladi levi, Ljubljana 2006.

Chapter 5, Samuel Thulin

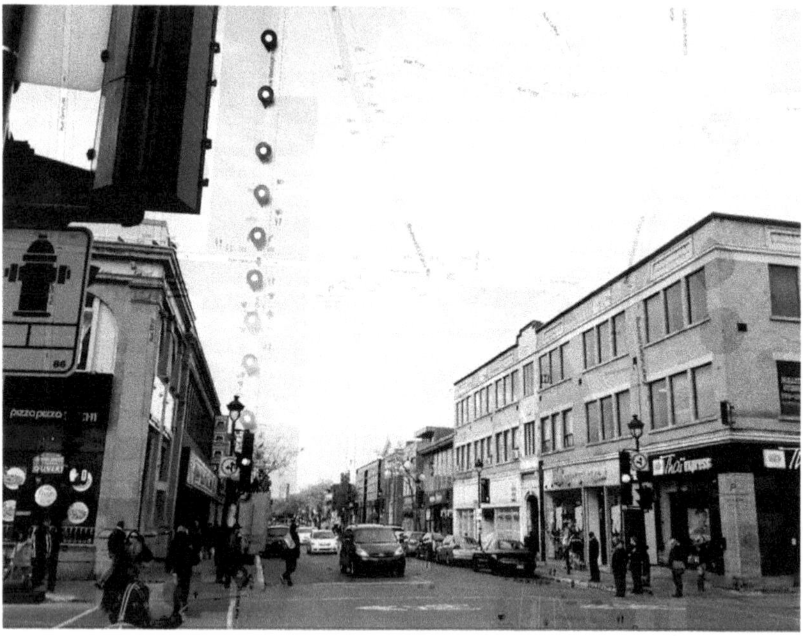

Figure 5.1 Overlay of street view and map view of Wellington Street in Montreal, the location of the Verdun Music-Route

Source: Samuel Thulin + Map tiles © openstreetmap.org contributors.

Chapter 8, Mike Collier and Kevin Hannam

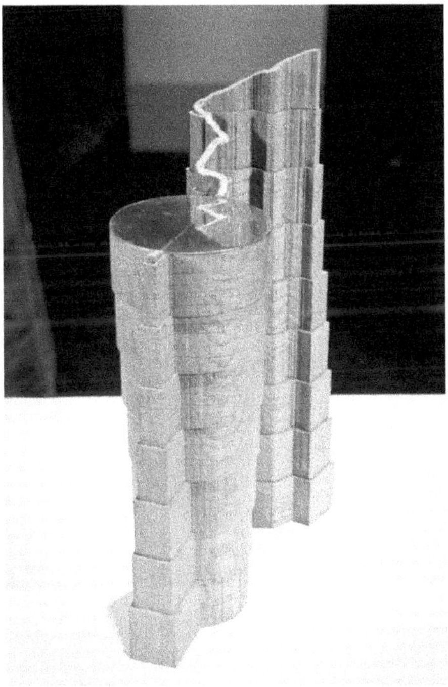

Figure 8.5 Brian Thompson, *Over the Ferry to the Station (with J, E & B)*, 2014, oak and copper, 29 x 17 x 8 cm

Source: Colin Davison—Rosella Studios.

Chapter 7, Nikki Pugh

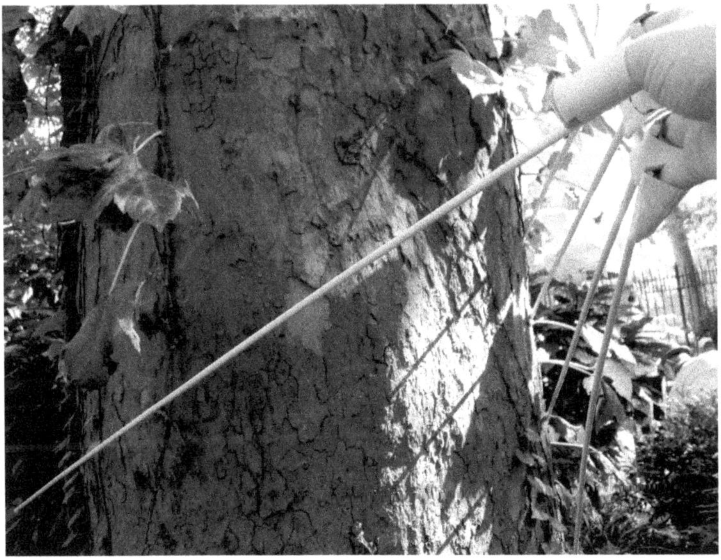

Figure 7.1 Exploring a mediated and expanded sense of touch

Source: Nikki Pugh.

Chapter 12, Lee Lee

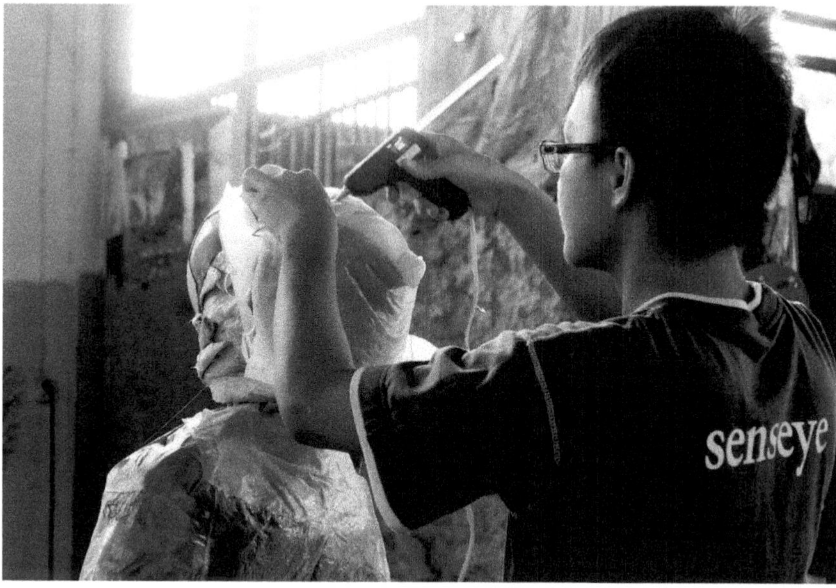

Figure 12.4 A student from the Stratton ABC Foundation uses repurposed plastic waste to construct the 'Plastic Demon' that migrated around Northern Thailand to raise awareness of plastic pollution

Source: John Cope, Stratton ABC.

Chapter 3, Jen Southern

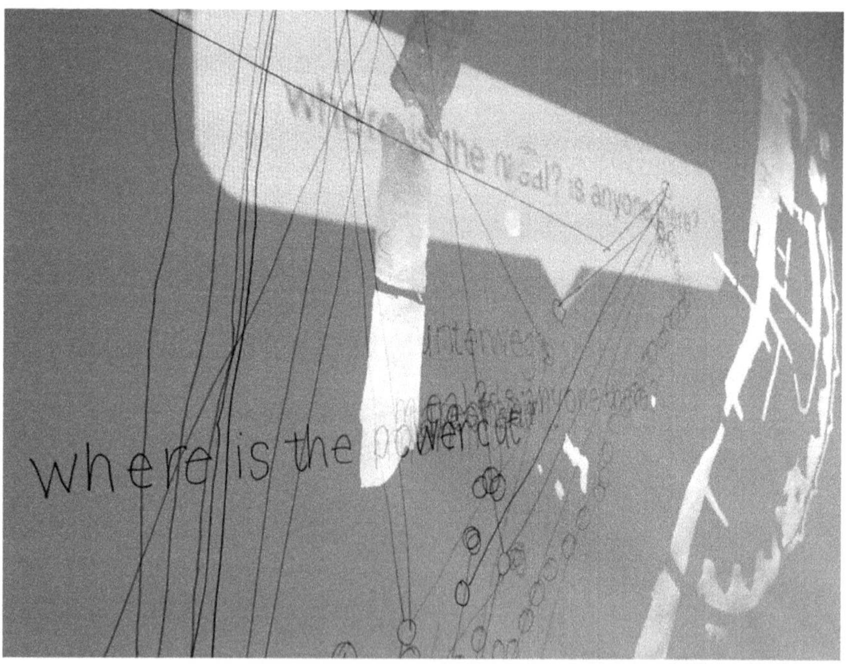

Figure 3.2 Detail of *Walking to Work No.3* at Networked Urban Mobilities 2014

Source: Jen Southern.

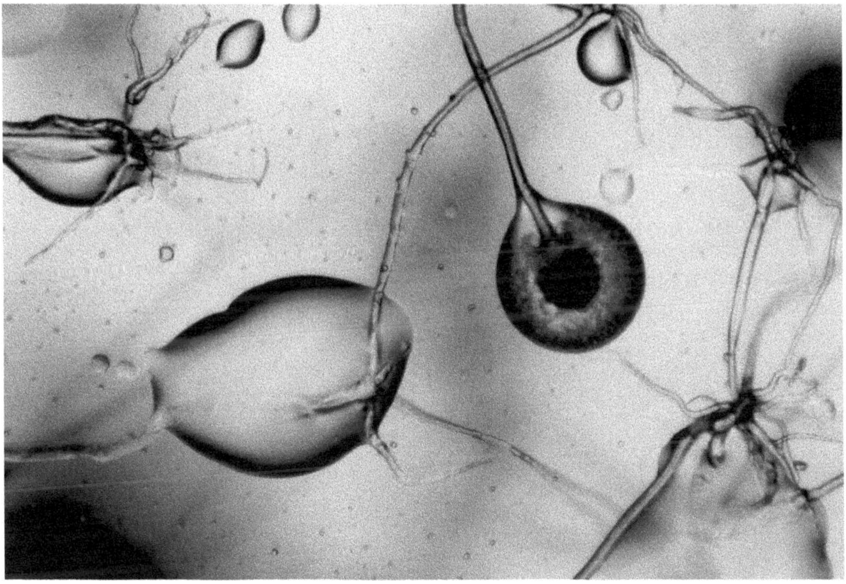

Figure 9.1 The Moldy Strategy. Domestic fungi captured with a digital microscope

Source: Antonia Hernández.

Chapter 9, Antonia Hernández

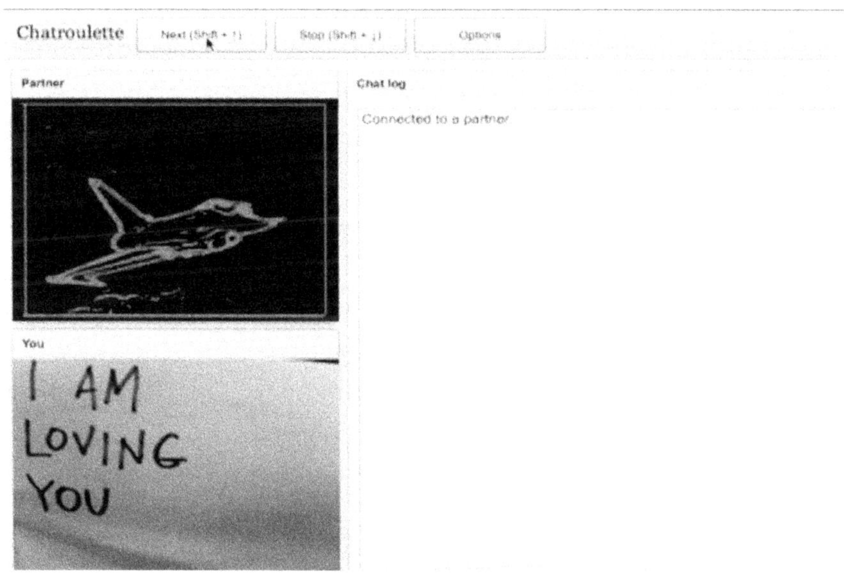

Figure 9.2 Performing Love: I'm Loving You. Screencapture of the online performance. The two video windows to the left show the connected participants. The main white window is intended for text chatting

Source: Antonia Hernández.

Chapter 14, Angie Cotte

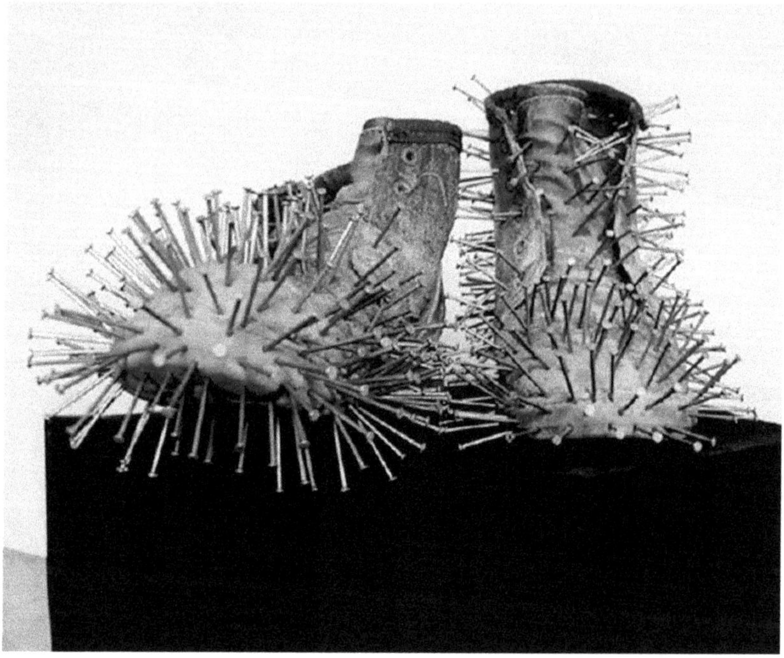

Figure 14.1 Illustration by Mohammed Al Hawajri—Cactus Borders project from 2005

Source: Permission from the artist.

Chapter 11, Ulrike Boskamp and Annette Kranen

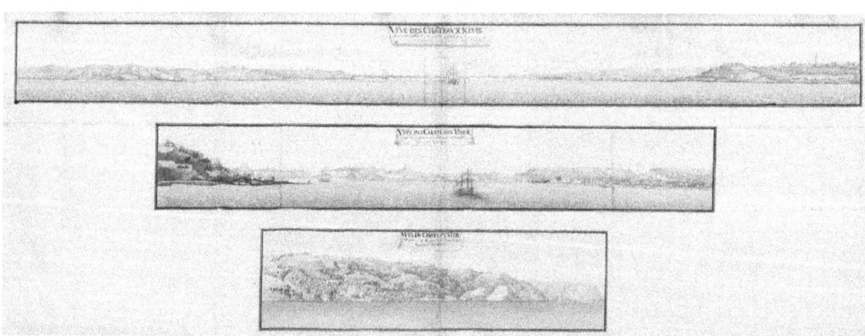

Figure 11.2 Étienne Plantier, Veüe des Châteaux Neufs des Dardanelles en venant de Tenedos; Veüe des Chateaux vieux des Dardanelles, dits Sestos et Abidos, avant de les passer; Veüe du Chateau vieil d'Europe, ou Sestos, prise devant la batterie de la mer, 1686, Paris, Vincennes, Service historique de la Marine

Source: © Service historique de la Défense, DBIB Vincennes SH 109.

9 Performative Fungal Strategies

Or How I Stopped Worrying and Started Loving the Network

Antonia Hernández

In the quiet room, the only light comes from the screen. It is not enough for the computer's tiny camera so I look as if I'm in an impressionist painting. *Next, next, next.* With each click I can find someone else. Next. This *someone* is generally alone, in another room, in front of a computer screen. Next. Looking for something. Next. As am I. Next, next, next. Video-roulette sites allow random webcam encounters among people unknown to each other on the Internet. You can *skip* your potential partner. They can do the same to you. Next. A group of people on a couch. Next. A young woman that knows she has better options than me. Next. Within a couple of seconds your partner or you will decide if the other deserves your time. Next. A shirtless person. Next. A dog, a child too young to be there, a masked man, my own image duplicated, a man playing his keyboard, more bodies. Next. If both randomly connected partners decided to stay, they can interact through image, sound, and/or text messages. Next. A living-room without visible humans. Next. Because in this environment the difference between a remote human partner and a pre-recorded one is crucial but sometimes difficult to establish, the first question will generally be directed to verifying if the partner is *real* or *fake*. Usually through a familiar wave of the hand. Next. A man in a darker room. Next. The communication will continue until one of the partners, by agreement or not, ends the encounter. *Next.* The only light in the room comes from the screen.

How I Stopped Worrying and Started Loving the Network is a series of art experiments that interrogate what we are doing when we are living with a digital network. However different, these pieces are speculative explorations of the home ecology and the affective qualities of media in the domestic environment. Commonly overlooked in mobilities studies, conducting studies in domestic space is an opportunity to reflect on the complex interaction and companionship among all sort of components, including in this case software and non-human bodies.

How I Stopped Worrying and Started Loving the Network uses my art practice as a method of inquiry into the nature of the domestic media ecology and online relationships within this context. It is also an attempt to communicate those findings as experience. This chapter includes thoughts and stills from two video pieces in this series: *The Moldy Strategy* (Figure 9.1) and *Performing Love: I'm Loving You* (Figure 9.2). Through a recorded performance on a video-roulette website, *Performing Love: I'm Loving You* asks about how feelings can be transmitted through a digital network and their materiality of them. In *The Moldy Strategy* the scale has been changed and a common inhabitant of the house, mold, performs under a microscope, being the vehicle for thinking through a network.

There Is No Place Like Home

The first question I put to myself was about the home. Or conversely, as Joseph Grima puts it: "home is the answer, but what is the question?" (2014: 12). My inquiry was more precisely about digital networks and domestic space, about the Internet and the home. As a recent immigrant, I wanted to know what it meant to be *at home*, and what the role of non-human elements and communication technologies was in that feeling. As Manuel de Landa (1995: n.p.) recalls:

> We tend to forget not only the flow of food but also the flow of electricity into our homes, as well as the electrical and hormonal flows in our bodies which play such crucial roles in the 'feeling of home'. And we tend to talk of the 'information age' without realizing that the future is as much about energy and materials as it is about information.

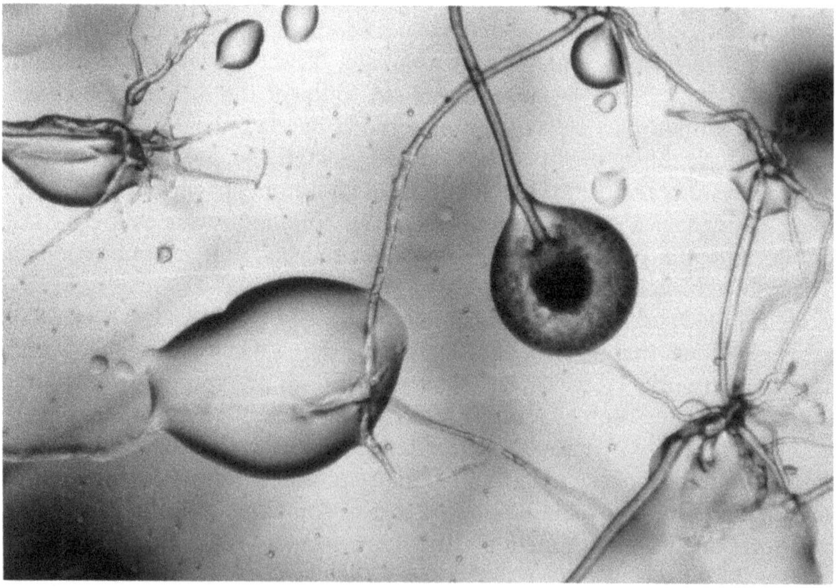

Figure 9.1 The Moldy Strategy. Domestic fungi captured with a digital microscope
Source: Antonia Hernández.

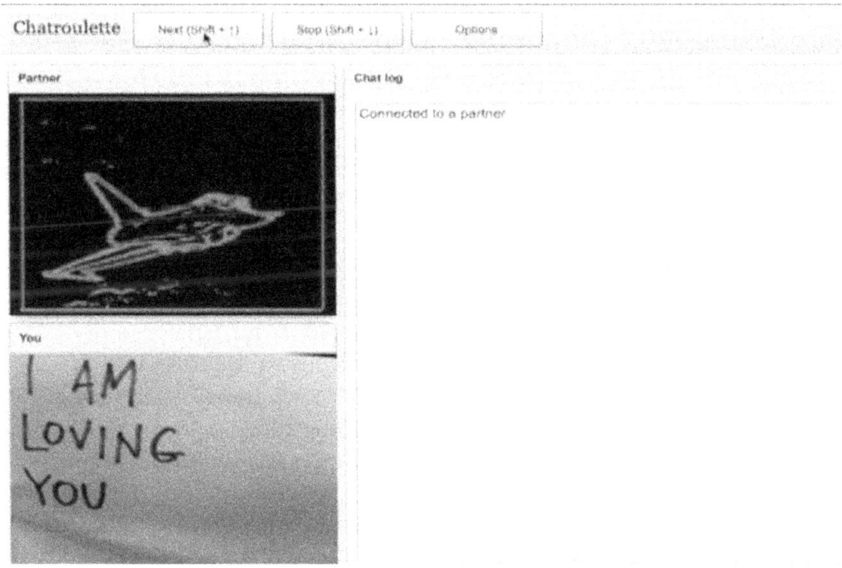

Figure 9.2 *Performing Love: I'm Loving You.* Screencapture of the online performance. The two video windows to the left show the connected participants. The main white window is intended for text chatting

Source: Antonia Hernández.

Where was my *being at home* happening? In the mundane conversations I was having with my mother in another country or in this place where I had no memories, no previous connections? Those conversations were not only real while they were happening: They impacted my environment, my habits, my relationships. At the same time, the new place I was inhabiting had an unfamiliar smell, things that belonged to other people, trajectories not fully contracted.

Even the distinction between *home* and *house* was new for me. Although there are similar words in Spanish, my primary language, their use is completely different. Soon I learned that *a house is not a home* but *a home is where the heart is*. Would it still be possible to consider this *home* in material terms, with media as part of the environment? If media can be understood as part of "an atmospheric grid of connections where distinct milieus adapt together" (Parisi, 2009: 182), creating "intensive environmental relations" (Parikka, 2010: 169), I wanted to know how online relationships were happening in this context.

An online interaction with family, a transposed idea of home, the role of communicating media in those configurations, seemed to be safe routes to explore. However, how do we account for the impact of screen-mediated communications with unknown people? With unknown unknowns? I started to explore video-roulette websites, which were at the time of these explorations (2011) still in early states of emergence. On those sites people are connected randomly, in real time, using webcams and text chat. Unlike exchanges with known or predetermined people, where the medium is secondary to the relationship, video-roulettes websites offered a brief opaque moment into the common transparency of digital media. At the beginning, a quick scan in these websites seemed to show the fascinating portrait of a new generation of Internet users: young, alone in their bedrooms, permanently connected, bored: the local face of a global phenomenon. However, despite my enthusiasm, the interactions with people through these sites were surprisingly predictable, with idiosyncrasies arguably inherited from mainstream pornography and common biases related to gender, age, and beauty standards. While the main *attractor* in video-roulette websites was presumably desire and the opportunity of an erotic interaction—or at least that was my unsupported vision—I was intrigued by the range of different expressions in that context. Was it possible, for example, to *love* unknown people? To *perform* love through a computer screen toward random people? And to receive it? As Luciana Parisi argues, technical changes are "inseparable from changes in the material, cognitive and affective capacities of a body to feel" (2009: 182) and I wanted to explore those mobilizations of affect and communicate those experiences as experience.

I developed a performative experiment: On my side of the screen I was behind a handmade sign saying 'I'm loving you' (Figure 9.3)—in an attempt to avoid common biases related to gender or age—trying to feel and express

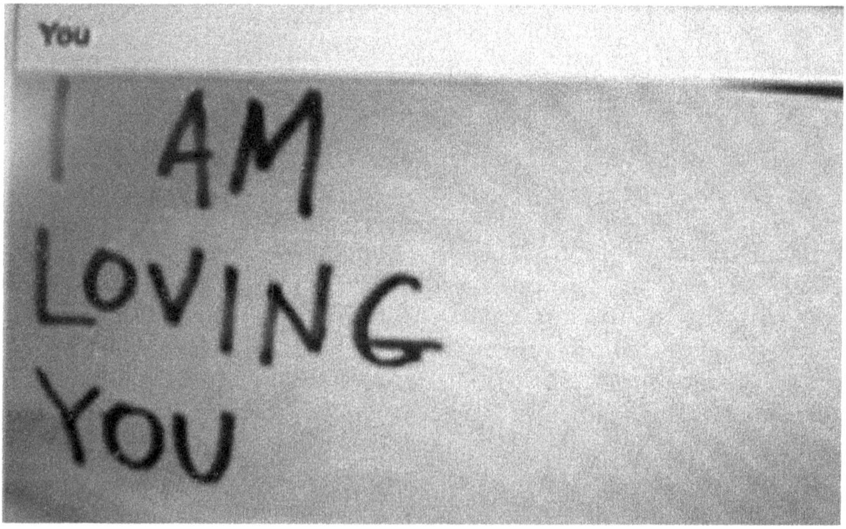

Figure 9.3 Performing Love: I'm Loving You. Screencapture of the online performance

Source: Antonia Hernández.

that emotion. I invited the random people the video-roulette connected me with to be loved and observed the response. The invitation was received with disconcertion most of the time, some curiosity, indifference, and anger. There was also a contact with someone broadcasting a distorted image of a military flight. But suddenly, in a connection with someone that did not show their face, something happened. There was a minimal conversation through the chat, and it felt, surprisingly, like love. I was recording the whole performance and I made a video with extracts of it, editing it with a soundtrack that aimed to help communicate the experience.

Companion Species

Under the assumption that domestic space was a complex ecosystem, I wondered about the effects that interacting with a digital network, such as the Internet, produces there: how is that experience affecting myself and my environment but also how does my behavior affect this network. I returned to the idea of the media ecology of the home, trying to consider the material dimension of digital communications (Terranova, 2004). Applying physical laws to live entities, following Erwin Schrödinger (1968), it is possible to affirm that a live entity (or a system) can endure while it succeeds in staying out of total and final equilibrium (entropy), feeding upon negative entropy. Using these considerations with the idea of home, this can be regarded as a system that evades entropy (final equilibrium or decay) by consuming negative entropy (order) from its environment, returning entropy to it: water that comes in clean and goes out dirty; vegetables that become garbage. While the cycle from food to waste is a clear example of this affirmation, what happens with communications? From where is the order taken and how is the waste returned?

It is helpful to borrow the question posed by Michel de Certeau (1988: 31), in which he asks about people and TV: "*what do they make of what they 'absorb,' receive, and pay for? What do they do with it?*" What was happening while communicating with unknown people through a digital network? An interesting complement to this idea is what is proposed by Robert Mitchell (2010), considering the word *media* not only in its common meaning, as channel, but in its biological sense of *medium*, a nutritive substrate. Following this distinction, online communications can be seen as expressed through media and as happening *in* media. Reversely, if the home is the place of these interactions, it should be considered their medium as well, where domestic space can act as a media itself. This aspect emphasizes what Mitchell calls the "affective history" of media, "a history that sees affect as a *generative* aspect of the processes of transformation that occur when new media emerge" (p. 93). Online interactions therefore—even with random people—seem to be an opportunity to think about media in generative terms, while home is affected by this event. At the same time, home can be experienced as an active medium and agent of the interactions that happen there. But how to account for this experience? I did not have the language for describing it. In what Anna Munster (2013) defines as 'network anesthesia,' there is an homogeneous and limited language for describing any kind of network, always frozen in time and space. Still exploring video-roulette websites, I did different attempts to *re-biologize* the people I encountered there, but I was not satisfied with the results. My aim was to avoid a representational analysis but I was in the difficult position of working with entities that I was unable to see or imagine at that point. However, if the problem was a product of human constraints, perhaps it could be overcome through non-human agents. With that hypothesis, I returned my attention back to the domestic ecology.

The Moldy Strategy

The constant need for cleaning made me suspicious of the human-centered order of the house. That order appeared not only as one of several possibilities but one that requires a constant battle against decay, through repeated normative acts. This decay involved entropy—or the tendency of things to go into disorder—and the action of microorganisms such as bacteria or mold. Having found rotten, moldy food in my refrigerator by chance (the kind of chance acquired by leaving food without supervision for a long period) I started to document it through photography, looking for different patterns of relations within the home. I was in a sense 'queering' my own domesticity, looking for those inherited performative acts that create reality there. Despite how alluring the pictures were, my lack of knowledge about mold limited my insight.

I started to collect mold, to cultivate and examine it through a microscope (Figure 9.4). While I expected some meshwork, what I found was even more captivating than the previous experience with the mold in the refrigerator. Embarrassed, I soon learned that indoor mold is actually a generic name for diverse kinds of microfungi that grow in domestic space (Flannigan, Samson, & Miller, 2001). Although some species are potentially pathogenic for humans, the common household co-exists with fungi in a harmless and sometimes symbiotic way (Zock et al., 2002). Fungi not only seduces me: Due to the variety of resources that fungi can use as a source of food, "it is difficult to think of any substance that they are unable to exploit or any habitat where they are completely absent" (Stephenson, 2010: 22). There are more than one million fungal species, while more than 1000 new ones are documented each year. Fungi owe their success to obtaining food from dead organic matter, growing through a complex and decentralized network, and by reproducing both sexually and asexually. The wonder that I saw through the microscope was not only beauty but an indefatigable network able to grow in creative and wise ways. It is not only about growth: It is about branching, fusion, and regression (Fricker et al., 2008). A network always in movement, where the environment is an active agent. Moreover, by analyzing more things found during the cleaning (a ball of hair from

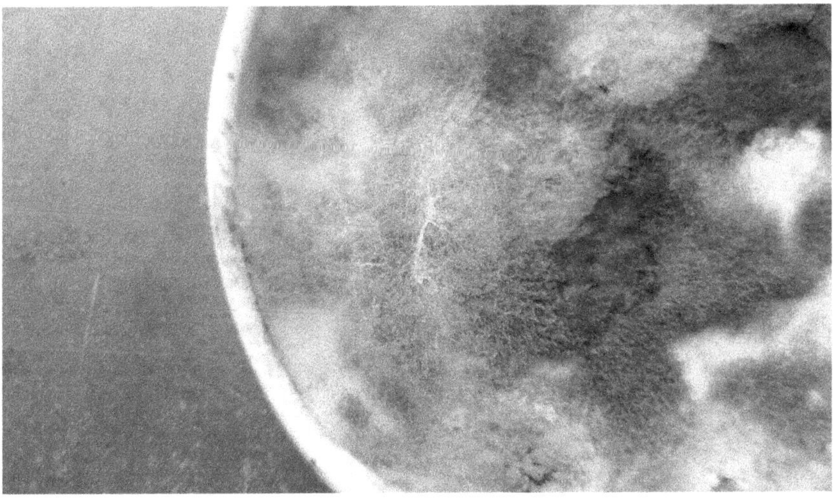

Figure 9.4 Rotten food found in the fridge
Source: Antonia Hernández.

the broom, a dead centipede) I realized that mold was present everywhere, waiting for an opportunity to grow, more active than any other inhabitant of the home. Furthermore, if in a relationship through computer screens different environments are put in contact, then these microfungi are also part of the experience and probably the clue to a different approach due their networked behavior. If I was looking for an alternative domesticity, a queer one, the fungi were doing something beyond my imagination. But 'queering,' as Jeffrey J. Cohen reminds us, "is at its heart a process of wonder" (quoted in Giffney & Hird, 2008: xxiv). Although I was far from actualizing the virtuality of online relationships, possibilities seemed to be expanded.

The new challenge of the project was to work with this exciting discovery: Mold is affecting me, I can affect it, we are entangled in this environment. Were human-fungal relationships a useful model for thinking through the experience of co-existence with a domestic-digital network? I was trying to avoid just using a new set of biological metaphors to apply to communications studies (Munster & Lovink, 2005) and instead find a way to interrogate and express what living with a digital network entails. A way able to consider the material experience of the home and the liveliness of networks.

Using a specialized microscope (Figure 9.5), I video-recorded fungi that I found in my home at different stages of development. Although at some levels of magnification it was possible to observe the displacement of food within the fungal network and the continuous spread of the spores, most of the movement achieved was the product of changes in the focus or

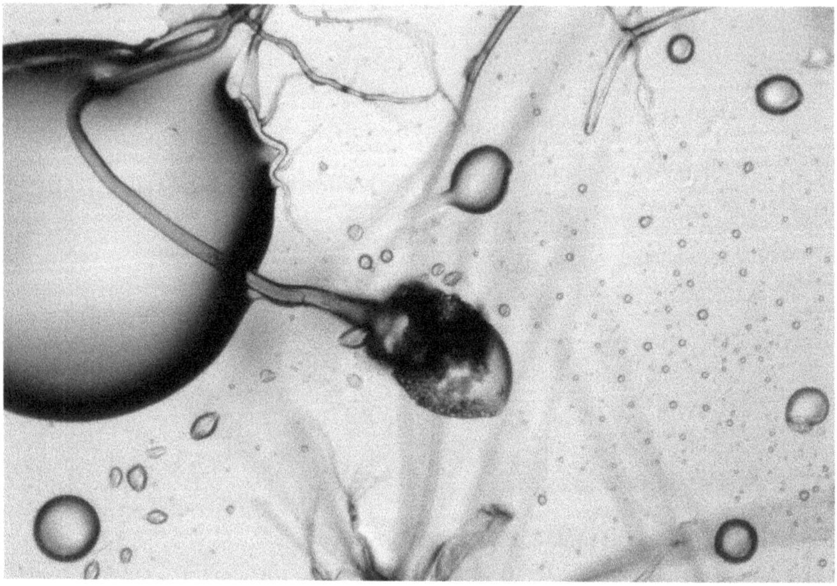

Figure 9.5 The Moldy Strategy. Domestic fungi captured with a digital microscope
Source: Antonia Hernández.

relocations of the object in the microscope by hand, achieving a sort of corporeal relation with the material. From several hours of recording, I made a film 40' in length. The soundtrack comes from a recording of a sexual encounter, magnified in time as the images are magnified by the microscope. I streamed the video through a video-roulette website instead of my own image, recording the reactions of my unknown partners. Those responses were surprisingly favorable, in terms of time devoted to see the image and questions about it. People were captivated by fungi just as I was. If what matters is to interact with a network, fungi seem to know more than any other user.

For the public display of this piece, the video has been projected in a large scale, trying to communicate the sense of being part of this biodigital medium. One of the objectives of this presentation was to reveal part of the complex ecology and composition of domestic space, where this network—now digitized—is acting. The space of the exhibition is activated by the action both of the fungi and people that circulated there. I was looking for an awareness about the affective relation with the environment, a feeling of being involved in this fungal network, and an experience of the media as a medium. The scale of the video and the soundtrack create the sense of being part of this medium, or "becoming-a-medium" (Mitchell, 2010: 99). Here, as it was outlined before, the idea of media is understood both as a channel and as a nutritious environment; a media that facilitates "a kind of communication that allows both new environments and new beings to come to existence" (p. 103). This innovative capacity, or generative one, expresses a vital quality that this project intends to embrace: the quality of affecting and being affected by the surrounding (Figure 9.6). This quality is arguably also present in the interaction with digital media (as in the case of the installation discussed here), not only due to a possible visceral reaction to the exhibition but from the capacity of the digital to manage and build up (new) media objects from discrete units and convert content from one platform to another. This conception of media can be extrapolated to networks, or the network, and is an opportunity to think of media as an assemblage (Parikka, 2010). *The Moldy Strategy* proposes to observe a network as a circulation of affect: A network that, rather than being constrained to a computer screen, involves the domestic space and the body of their inhabitants—all of them.

The Moldy Strategy has been an exercise in thinking through the speculative but possible encounter between mold and digital networks, challenging at the same time discursive formations of the home. The fungal model has given insights to the nature and behavior of a network, some clues about non-human domesticity, and how media is a realm of affects and densities. Mold appears here not as a metaphor for a network but as a vehicle that allows the imagination of some virtualities discarded in a human-based

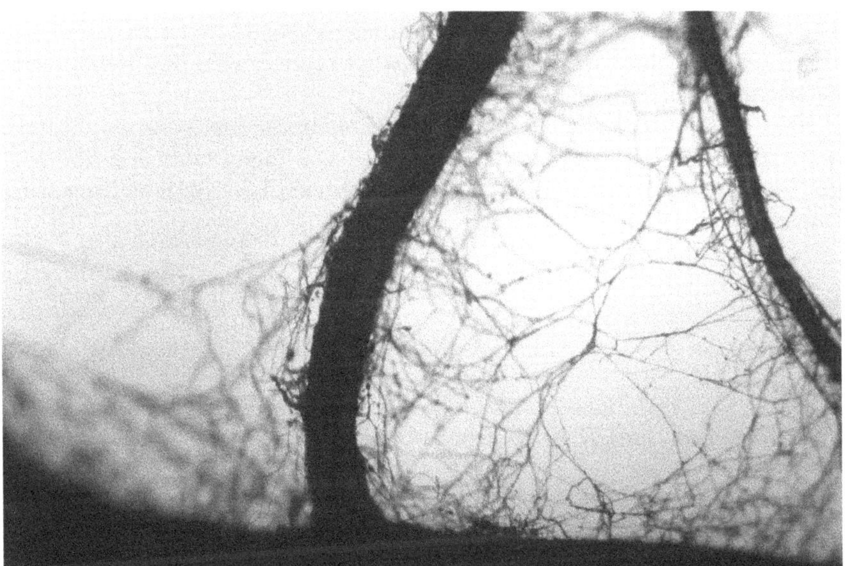

Figure 9.6 The Moldy Strategy. Domestic entity captured with a digital microscope

Source: Antonia Hernández.

model. By stressing the importance of non-human elements, an affective consideration of the ecology of domestic space shows how bits from different sorts can be mixed creating unexpected entities. An exploration of domesticity in a networked context offers an opportunity to reflect on the entangled relationship between bodies and media, physical and metaphorical language. This complex interaction presents an alternative perspective for reconsidering the currently uniform model and vocabulary used to describe any kind of network, while taking into account processes, influences, and relationships among their components.

Conclusion

I finished this project with more questions than I had at the beginning, questions that grew in a fungal way. However, *How I Stopped Worrying and Started Loving the Network* provided me an opportunity to reflect on and work through some concepts that I envisioned only on an abstract level before. More than facts, I learned about the possibilities that emerge from working with abstract concepts in a material way and then mobilizing them through art and practice. As Jussi Parikka affirms, practice also can be considered a "theoretical excavations into the world of 'things'" (Parikka, 2011, p.34). In this project, the theoretical research has been led/corroborated/modified by the artistic practice, which in turn has faced a similar process. I understood some of the theoretical concepts I was working with through material experimentation, which in turn led to abstract ideas. The process of thinking through non-human elements, such as fungi, creates a space in which we can consider reality beyond what has been already defined as possible by humans.

How I Stopped Worrying and Started Loving the Network would have never been possible without the support of Dr. Tagny Duff and FluxMedia, and Dr. Kim Sawchuk and the Mobile Media Lab, both at Concordia University.

References

De Certeau, M. (1988). *The Practice of Everyday Life*. Berkeley: University of California Press.

De Landa, M. (1995). Homes: Meshwork or Hierarchy? *Mediamatic Magazine*, *8*(2/3). Retrieved from http://www.mediamatic.net/en/page/8931/homes-meshwork-or-hierarchy

Flannigan, B., Samson, R. A., and Miller, J. D. (Eds.) (2001). *Microorganisms in Home and Indoor Work Environments: Diversity, Health Impacts, Investigation and Control*. Boca Raton, FL: CRC Press.

Fricker, M., Lee, J., Bebber, D., Tlalka, M., Hynes, J., Darrah, P., . . . Boddy, L. (2008). Imaging Complex Nutrient Dynamics in Mycelial Networks. *Journal of Microscopy*, 231(2), 317–331.

Giffney, N., and Hird, M. J. (2008). *Queering the Non/human*. Aldershot: Ashgate Publishing Co.

Grima, Joseph. (2014). Home Is the Answer, but What Is the Question? In Space Caviar and Biennale Interieur (Eds.), *SQM: The Quantified Home* [an Exploration of the Evolving Identity of the Home, from Utopian Experiment to the Factory of Data; on the Occasion of the 24th Biennale Interieur, October 17–26, 2014, Kortrijk, Belgium]. Zürich: Lars Muller Publishers, 2014.

Mitchell, R. (2010). *Bioart and the Vitality of Media*. Seattle: University of Washington Press.

Munster, A. (2013). *An Aesthesia of Networks: Conjunctive Experience in Art and Technology*. Cambridge, MA: MIT Press.

Munster, A., and Lovink, G. (2005). *Theses on Distributed Aesthetics: Or, What a Network Is Not*. Available at: http://mediacultures.net/jspui/handle/10002/627

Parikka, J. (2010). *Insect Media: An Archaeology of Animals and Technology*. Minneapolis: University of Minnesota Press.

Parikka, J. (2011). FCJ-116 Media Ecologies and Imaginary Media: Transversal Expansions, Contractions, and Foldings. *The Fibreculture Journal*, 17, 34–50.

Parisi, L. (2009). Technoecologies of Sensation. In B. Herzogenrath (Ed.), *Deleuze/Guattari & Ecology* (pp. 182–199). Hampshire: Palgrave Macmillan.

Schrödinger, E. (1968). *What Is Life?: The Physical Aspect of the Living Cell; with, Mind and Matter; & Autobiographical Sketches*. Cambridge and New York: Cambridge University Press.

Stephenson, S. (2010). *The Kingdom Fungi: The Biology of Mushrooms, Molds, and Lichens*. Portland, OR: Timber Press.

Terranova, T. (2004). *Network Culture: Politics for the Information Age*. London, Ann Arbor, MI: Pluto Press. Available at: www.loc.gov/catdir/toc/ecip0415/2004004513.html

Zock, J.-P., Jarvis, D., Luczynska, C., Sunyer, J., and Burney, P. (2002). Housing Characteristics, Reported Mold Exposure, and Asthma in the European Community Respiratory Health Survey. *Journal of Allergy and Clinical Immunology*, 110(2), 285–292. Available at: http://doi.org/10.1067/mai.2002.126383

10 Stop and Go

Investigating Nodes of Transformation and Transition

Michael Hieslmair and Michael Zinganel

The fall of the Iron Curtain was followed by a kind of bottom-up re-unification of Europe, which has gone largely unnoticed by scholars thus far. This re-unification, the German historian Karl Schlögel (2009) argues, was driven to a large extent by the many small-scale vendors who traveled the Pan-European corridors between the former East and West to purchase and resell goods on informal and formal markets, long before expensive infrastructure was implemented thanks to EU programs and large-scale logistics enterprises started to build their networks and hubs.

These nodes might be ephemeral, become obsolete, derelict, and substituted by others. Informal nodes can become formalized, normalized, and controlled while new nodes emerge at other places. They do not exist solely as single entities but rather as a part of a network of many different nodes, which are encountered during each individual's tour. Hence, they represent a new, more dynamic model of urbanity (Bittner, Hackenbroich, & Vöckler, 2006) constituted by interconnected polyrhythmic ensembles (Crang, 2001) or archipelagos.

The corridors that interconnect these nodes represent monuments of the modernization of nation states and supranational institutions, rich sources of imagination that trigger dreams (and nightmares) of both institutions and individuals. These corridors operate like magnets (Stewart, 2014: 552), attracting both objects and people to move on them or to agglomerate alongside, "doing with space" (Lussault & Stock, 2010: 12) in a variety of ways. What is experienced and encountered here is registered in statistical records of controlling and policing institutions, in newsreels of mass media, in campfire narratives, and in everyday life myths of road users and road neighbors—and also in the research reports of scholars.

In a project like ours research destinations and routes are first selected to support the author's theoretical arguments, largely inspired by mobilities studies (e.g. Urry, 2007; Cresswell, 2010; Cresswell & Merriman, 2011). In our case the funding program of the City of Vienna initially predefined the primary research destination. Our perspective: By examining the changes

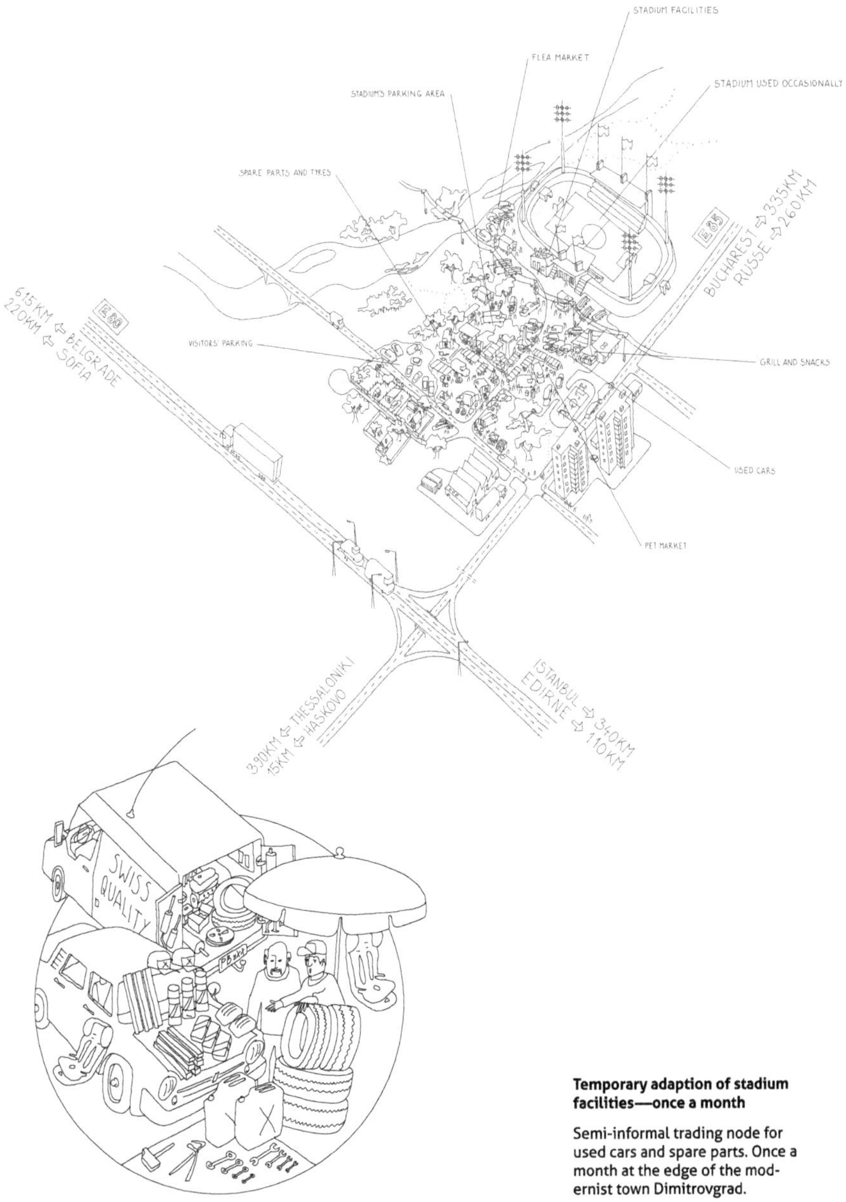

Temporary adaption of stadium facilities—once a month

Semi-informal trading node for used cars and spare parts. Once a month at the edge of the modernist town Dimitrovgrad.

Source for Figures 10.1 10.7 Michael Hieslmair | Michael Zinganel

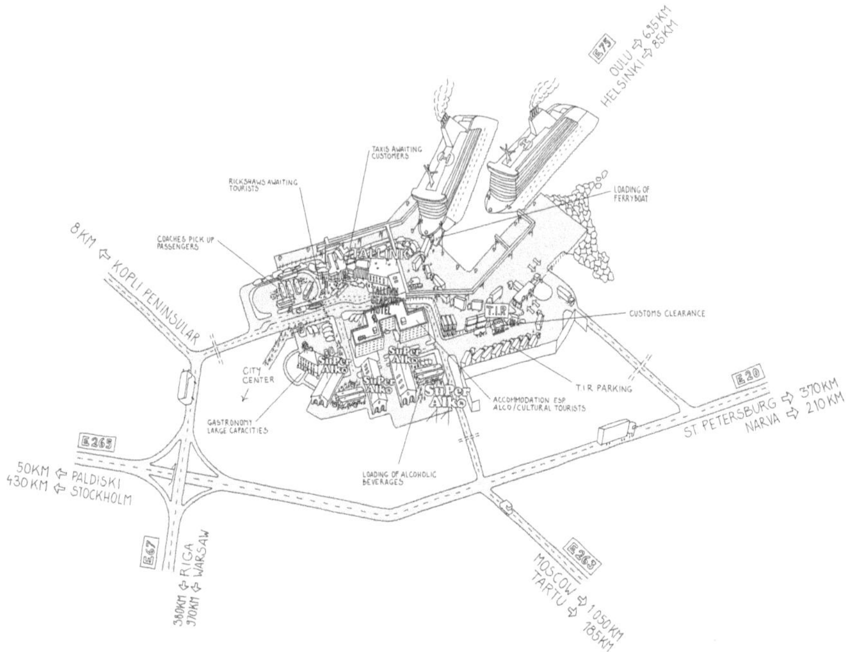

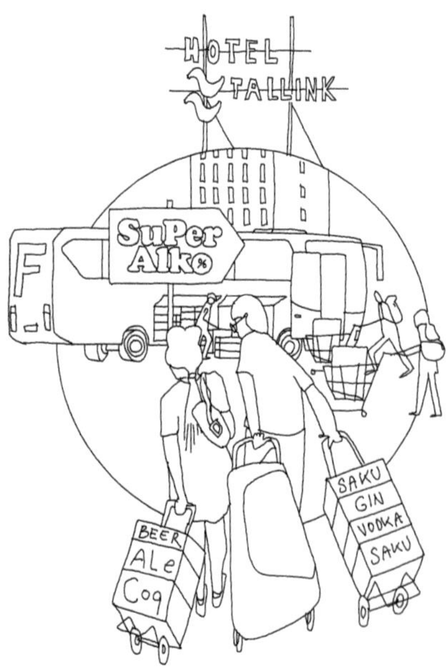

Various rhythms affecting the harbor area—permanent

Tallinn harbor as a polyrhythmic ensemble for different modes of transportation and cross-border activities.

in transnational mobility patterns in Vienna, a capital of the 'former West,' the post-socialist transformation of the wider geopolitical sphere—the former East of Europe—can be easily investigated as well. But how to get on the road? To this end, we called for a comparative approach and a multi-sited research and chose two additional destinations: Tallinn in Estonia and the Bulgarian–Turkish border region, located at the two opposite ends of an important north–south axis, which are characterized by quite different geopolitical constellations and tensions, both historically and in contemporary political transitions and urban transformations. The two regions also differ significantly in terms of state regulation, policy, and in the quantity and quality of design. But later, during our field trips, certain places would attract us for other, often more convincing, reasons.

Method 1: Embedded Research: Attempts to Assimilate Into the Field of Research

In keeping with the design of our study, the core elements of investigation should be at least two intensive research trips along the geographic triangle of Vienna, Tallinn, and the Bulgarian–Turkish border. Referring to 'mobile methods' (Büscher & Urry, 2009), we proposed a mobile ethnography to become physically immersed in mobile activities while simultaneously working on material and visual representations of networked mobilities.

We therefore purchased a cheap trailer, licensed to drive on roads for transporting small boats, which we transformed into a mobile display for drawing large-scale maps of the travel experiences of the mobile actors we expected to meet at the stops along our routes. To pull the trailer on our three-week trips we originally wanted to rent a small van. However, we came to learn that there was no chance of renting any vehicle to drive to high-risk areas like Serbia or Bulgaria due to potential car theft. So we needed to purchase a second-hand car ourselves. But in the entire metropolitan area of Vienna all second-hand cars suited for fitters and vendors and available for a reasonable price had already been sold or reserved for car dealers from Eastern Europe. Even when we expanded our search radius to more than 150 km we failed. Coincidentally, we passed by a rural gas station, where several Ford Transit vans were parked. The owner, a motocross racing driver who was in the hospital after a bad accident, urgently needed to sell one of the vans to refinance spare parts for his bike. We took our chance! When we returned a few days later to pick up the van, we learned that several Serbian car vendors had already asked for it. Our attempt to assimilate into the social group of vendors was proving successful, even before we started our official research trip.

Small transporters like ours are an attractive choice to escape the strict regulations and control mechanisms imposed on drivers of full-size trucks. They offer the opportunity to drive seven days a week around the clock and bypass the long waiting queues of TIR border controls to transport goods and people—with or without proper papers.

The decision to purchase a car also had a strong impact on the design of the study: We were able to drive more often and be much more flexible according to the routes and rhythms of our trips. It also shifted our interests: the significant amount of private property we had invested in should return one day in the future. Our plan was to resell the vehicle at the end of the project, exporting it and offering it for sale at one of the many second-hand car markets along our routes—and thereby also re-integrating it into the economic cycles of our research destinations (moreover, tracing the future mobility career of our Ford Transit would also be nice follow-up project to our research).

Already during our first trip from end of July to mid of August 2014 we visited the car market of Marampolje in Lithuania close to the Polish border, famously described by Karl Schlögel in support of his thesis (2005). Starting immediately after the fall of the Iron Curtain, second-hand cars from Western Europe were brought to a big parking lot in front of a vacant factory; some

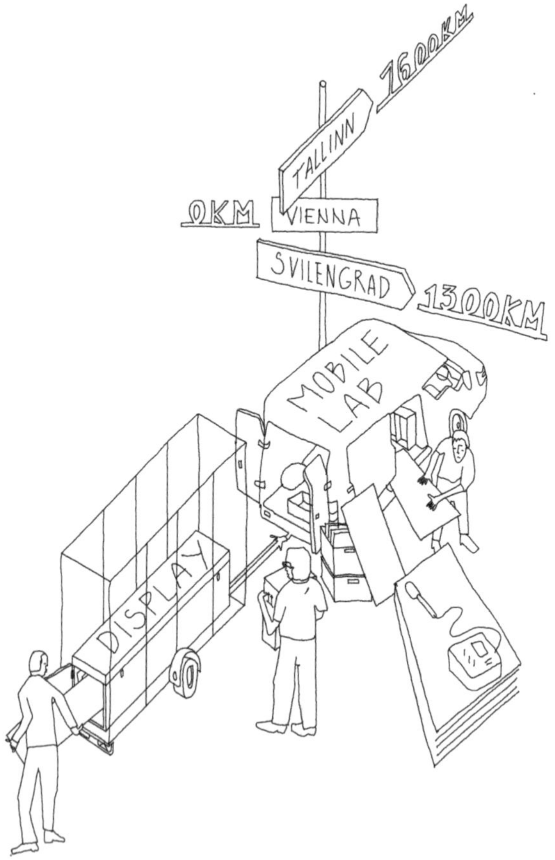

Packing the mobile lab and preparations for the research trips

A Ford Transit transporter and trailer with audiovisual equipment for documentation and tools for building up on-site interventions. Heading for places in a triangle between Vienna–Tallinn, and the Bulgarian–Turkish border.

were sold directly from big trailers to end users on site, others to vendors that transported them to post-Soviet states reaching from Kaliningrad and the Baltic to the Caspian Sea. The market became increasingly formalized—in contrast to a much smaller market in Dimitrovgrad in Bulgaria we visited later: Here, close to the crossroads of several highway corridors, the special rhythm of the market imposes a substantial dynamic on the place. The market is located in a park landscape around a derelict football stadium, offering parking lots, toilets, kiosks, and grill areas. The market only happens on the first Sunday of each month and is open just from early morning till noon. The lacking continuity of the offers, the limited time to stroll around, to enjoy the visual, social, and culinary diversity, and eventually to purchase an item heightens the visitors' curiosity and demand. The short time slot also prevents vendors from investing in more solid or even permanent stalls. The products offered range from fully functional cars—many of them with number plates from Great Britain—to spare parts and all kinds of small-scale technical items, such as vacuum cleaners and music CDs, etc., which are displayed for sale around the cars, in the trunks, and on their hoods.

Method 2: Mapping on Route: Visual Representations and Tools for Interaction

Following Deleuze's (Deleuze & Guattari, 1987) and Latour's (2005) enthusiasm, we argue to view maps not only as a proper device for the

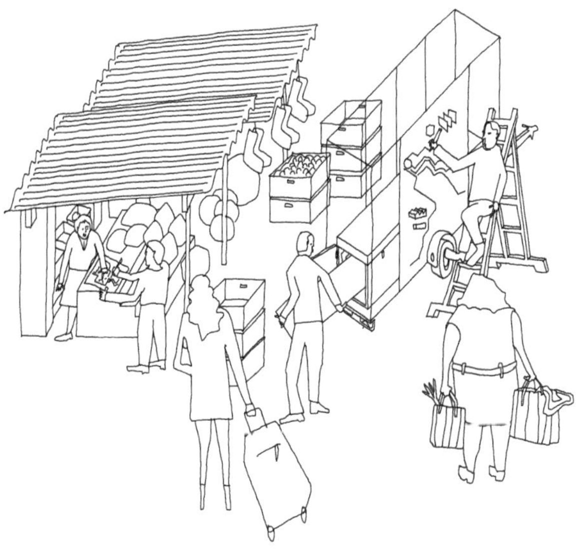

Live drawing during episodic interviews

Trailer transformed into a large-scale drawing board and display for mapping exercises in the research areas. Live drawing triggers conversations with passersby.

representation of mobility patterns but moreover live mapping as a great tool for the stimulation of interaction with mobile actors en route. We first tested this method at a market in Tallinn, where predominantly ethnic Russians offer food, cheap clothes, and Soviet souvenirs. We parked our van and trailer at a place between the main entrance and the adjacent railway station, pretending to unload some stuff. After mounting white panels to the trailer, transforming it into a large-scale drawing board, we wanted to ask visitors and especially vendors about their own migration history (from then Soviet territories), their everyday routes from their home to the market, and where and how they get their goods to sell. But since vendors cannot leave their stalls during market hours, we had to change the strategy: We walked to the stalls to draw sketches on paper first and then transferred them onto our oversized drawing board. The rather abstract lines representing the paths how our interviewees drove through the city were interrupted by comic style sketches of significant buildings and by the means of transport used by them. And sure enough, the fragmentary diagram immediately attracted other people to tell us their own stories.

In contrast to our naive expectations, most of the cheap goods offered by the Russian vendors had not been imported from Russia, but rather from a market south of Warsaw, beside the route we drove before. We decided to visit this market upon the next occasion we came through: It turned out to be the largest wholesale market for Asian products in Europe, run by Turkish, Vietnamese, and Chinese immigrants.

Method 3: Interventions in Public Space: Material Representations of Mobile Phenomena

In September 2015 we realized our first large-scale intervention in the public space of the Tallinn harbor area. An obvious node of transnational mobility and migration with a high impact on the rhythm and economy of the city of Tallinn, it is strangely outside of the focus of Tallinn-based scholars and artists. We employed the cultural capital of the Tallinn Architecture Biennale, where we had invited ourselves, to convince the harbor administration to support our project.

We obtained permission to park our trailer in front of the highly frequented Terminal D. In this context the trailer now functioned as the supporting structure for a three-dimensional network sculpture. The yellow wooden beams represented an abstract map of the routes and paths of selected individuals that use the ferry connection between Tallinn and Helsinki. Via built-in loudspeakers passersby could listen to their narrations, which introduced different motives, rhythms, rituals, and routines: for example, of an ethnic Russian teacher who works in shifts at a bar on the ferry boat after having lost her original job; of a Russian business man

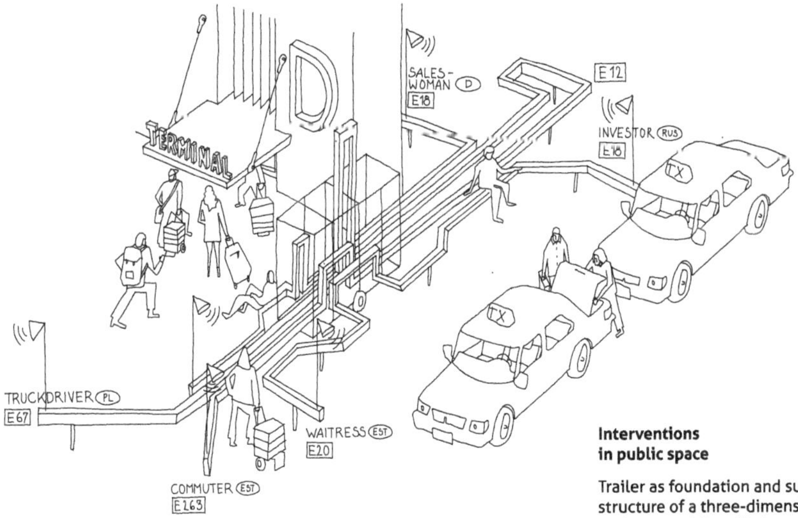

Interventions in public space

Trailer as foundation and substructure of a three-dimensional path-network installation equipped with loudspeakers in front of Tallinn's ferry terminal.

from St. Petersburg who checks his real estate investments in Estonia every second month; of a Polish truck driver who passes by the harbor once a week; of a construction worker from the southeast of Estonia who works as labor migrant around Helsinki; of a sales representative of a German company who visits clients in Scandinavia and the Baltic area; and of a group of young Finns who frequently travel to Tallinn with a reduced group ticket to stroll around the bars of the old town of Tallinn, before returning with their shopping carts to bring back as much alcoholic drinks as the carts allowed. The significant difference of prices easily enables them to refinance the costs of trip.

This installation was not only intended to be a representation of the final results of our Tallinn research but also a visually attractive trigger for collecting additional expertise from mobile actors. During the setup process we could experience live on site how the arrival times of ferry boats dramatically increased the number of taxis, buses, and rickshaw drivers along with the presence of small-scale vendors and beggars. The seemingly overdimensioned asphalt desert of parking lots became completely filled with vehicles for a limited period of time. Once the parking lots were empty again, the consumption zones alongside the beaten paths of tourists had become crowded: the souvenir shops and bars in the Old Town, the shopping malls along the inner ring, and the many alcohol supermarkets and hotels near the harbor.

Method 4: Academic and Artistic Workshops: From the Mobile to the Stationary Lab

As the van we had purchased was much larger than originally intended we were able to transport larger pre-produced maps, but we could also collect significant everyday life objects during our trips, additional 'actants' with reference to Latour's ANT (2005), and even art pieces—items we found relevant to combine with the materials from interviews, observations, and the artistic representations.

These objects were then displayed in the van but moreover also in interdisciplinary workshops and exhibitions in Tallinn and Sofia. In addition or juxtaposition to 'going native' we also invited both scholars and artists to make presentations and discuss the theories, methods, and interim findings of our own project, but also to expand the networks of experts on site as well. Guests were invited to present their projects as papers but also accompanied with exhibited objects, network diagrams, photo essays, and videos. All this material was also integrated into other presentations, especially at our own project space in Vienna. For this purpose, we had chosen the terrain of a former railway station relatively close to the city center, which is still used as a rail-to-road container and cargo terminal and as a hub for many small logistic companies—a place with a vivid relationship to the issues, a great atmosphere, lots of mobilities expertise, and certainly worth a future research project of its own. Here we worked on a networked cartography—or deep multimedia

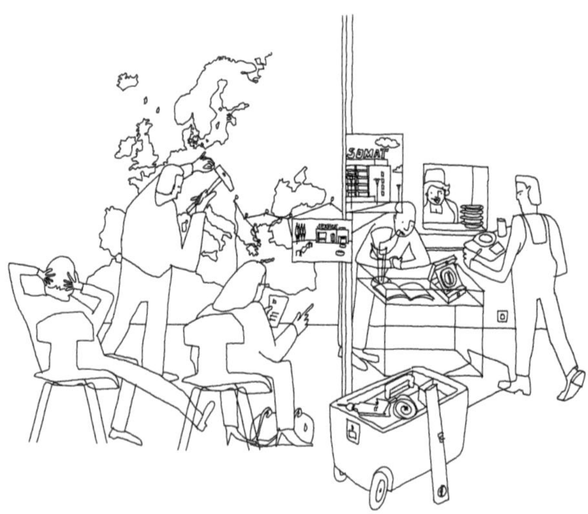

Workshop and exhibitions: on location, at our project space, in institutions

Circulation of different types of knowledge as temporary exhibitions and as conceived for interdisciplinary workshops in Tallinn, Vienna, and Sofia (e.g. intervention in a truck drivers' canteen at a former SO MAT base).

mapping—of our research trips and the research findings from our case studies, which will be continuously densified parallel to the progress of the project.

At this location we also host events like public workshops, lectures, and film presentations. For example, the final part of a workshop in December 2015 was a one-day bus trip to significant nodes of transnational mobilities and migration in and around Vienna: Starting at the international bus terminal we visited the container terminal in the Danube harbor, the logistics agglomeration at the airport, a designer outlet center at a highway crossroads, and—as a final attraction—the empty border station between Austria and Hungary in Nickelsdorf, which we had passed several times during our research trips before.

This last site had ultimately picked us. Originally, our aim was to investigate the normality of mobilities and migration. But after the dramatic increase of refugees in autumn 2015 had transformed the sleepy border village into a hotspot for migration management and international media coverage, we could no longer ignore it. Interestingly enough, none of the infrastructure facilities that the major guided us along were new: Within a matter of a few hours the former border station—deactivated yet not dismantled after the end of its service as the EU and later Schengen border—had been reactivated. Soon it had been supported by other auxiliary structures provided by emergency aid NGOs and event industries to accommodate an average of 4,500 (with a climax of 17,000) refugees per day in a village of only 1200. Buses, used for commuters and tourists thus far, transported refugees to the train station or to emergency shelters located in the vicinity of Vienna's bus terminals, where the local and/or federal government had access to immediately appropriate and adapted public buildings for their accommodation.

After Hungary decided to 'close' and control the Schengen border with Serbia and later with Croatia, the routes of the refugees had been redirected: The border station became defunct once again and life in the village boring as before. But this exciting interim state of emergency had not been unique in the history of mass migration in this little border village: 180,000 refugees already passed through the village during the Hungarian crisis in November 1956, and after the fall of the Iron Curtain in autumn 1989 40,000 exhausted GDR citizens had to be hospitalized here. And also the redirected routes for forced migration were not new at all: They changed to well-established trading and smuggling routes, where local experts can rely on the logistics and networks, including the staff of border control units. They drive the very same routes as millions of other mobile individuals, the very same routes we drove during our research project—and very often in the same type of car.

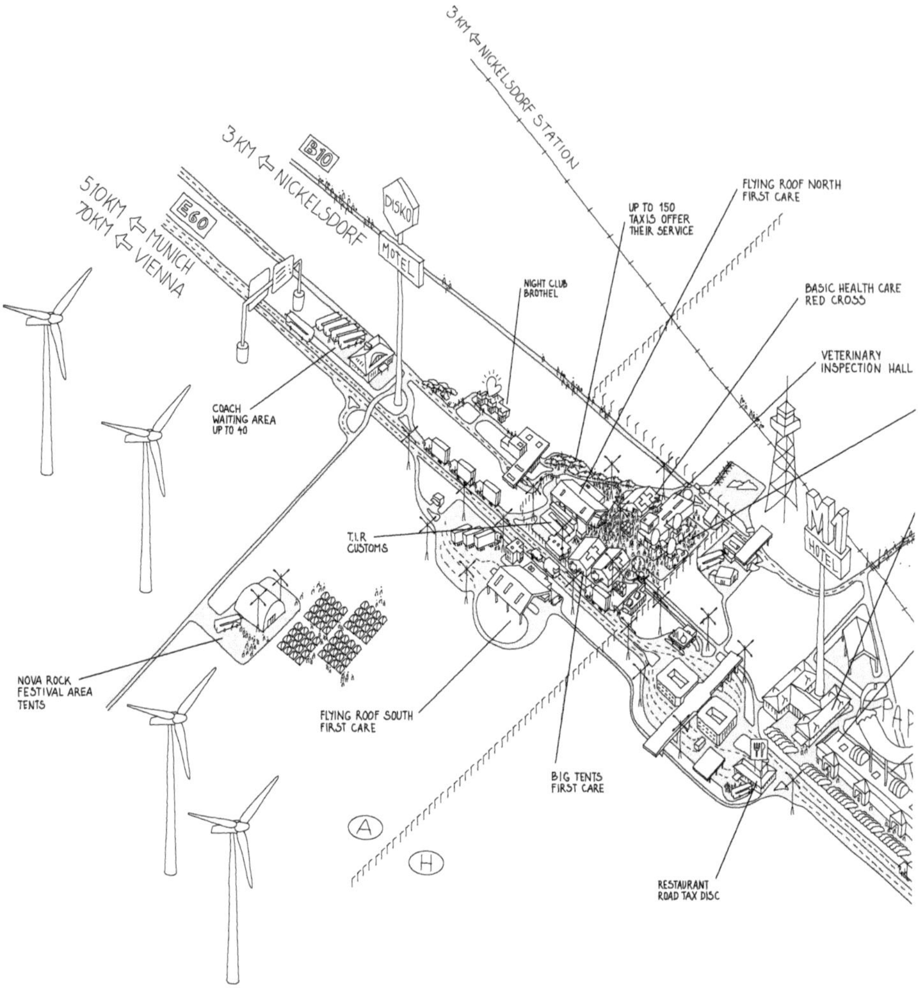
3KM ⇦ NICKELSDORF STATION
3KM ⇦ NICKELSDORF
B10
510KM ⇦ MUNICH
70KM ⇦ VIENNA
E60
DISKO
MOTEL
FLYING ROOF NORTH
FIRST CARE
UP TO 150
TAXIS OFFER
THEIR SERVICE
NIGHT CLUB
BROTHEL
BASIC HEALTH CARE
RED CROSS
VETERINARY
INSPECTION HALL
COACH
WAITING AREA
UP TO 40
M1
HOTEL
T.I.R
CUSTOMS
NOVA ROCK
FESTIVAL AREA
TENTS
FLYING ROOF SOUTH
FIRST CARE
BIG TENTS
FIRST CARE
A
H
RESTAURANT
ROAD TAX DISC

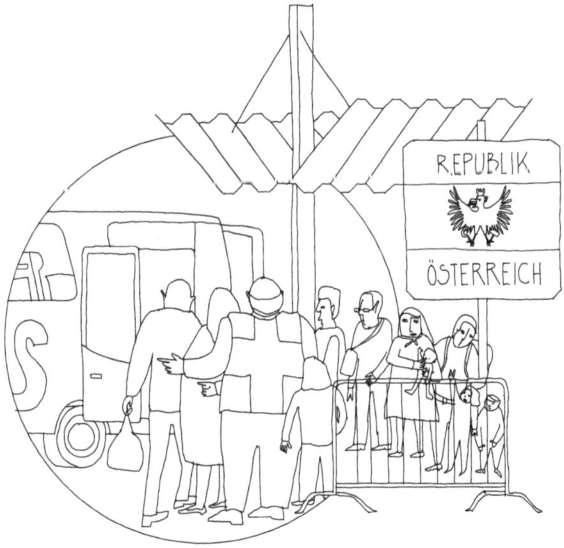

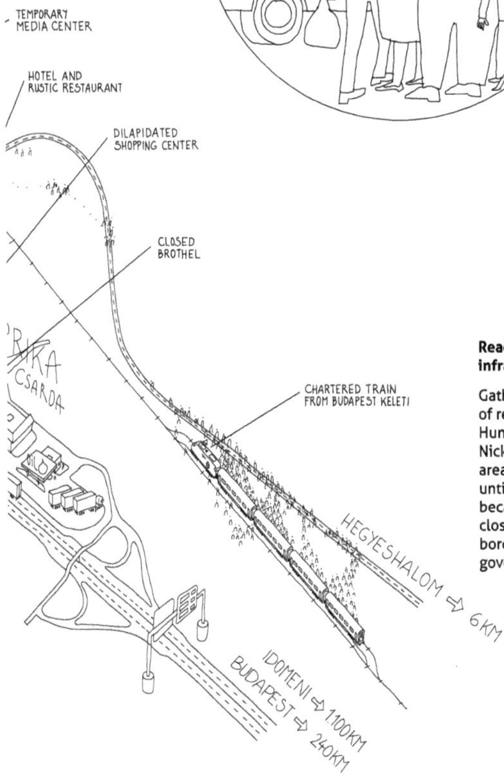

Reactivated border infrastructure—July to October 2015

Gathering area for a large number of refugees at the Austrian–Hungarian border checkpoint Nickelsdorf. First care and waiting area for peoples' onward transport until the main migration route became redirected due to the closure of the Serbian–Hungarian border initiated by the Hungarian government.

References

Bittner, R., Hackenbroich, W., and Vöckler, K. (Eds.) (2006). *Transiträume: Transit Spaces*: Edition Bauhaus, 19. Berlin: Jovis.

Büscher, M., and Urry, J. (2009). Mobile Methods and the Empirical. *European Journal of Social Theory* (February 2009), 12, 99–116.

Crang, M. (2001). Rhythms of the City: Temporalised Space and Motion. In J. May and N. Thrift (Eds.), *Timespace: Geographies of Temporality* (pp. 187–207). London: Routledge.

Cresswell, T. (2010). Towards a Politics of Mobility. *Environment and Planning D: Society and Space*, 28(1), 17–31.

Cresswell, T., and Merriman, P. (Eds.) (2011). *Geographies of Mobilities: Practices, Spaces, Subjects*. Farnham and Burlington, VT: Ashgate.

Deleuze, G., and Guattari, F. (1987). *A Thousand Plateaus: Capitalism and Schizophrenia*. (B. Massumi, Trans. and Foreword). Minneapolis: University of Minnesota Press.

Latour, B. (2005). *Reassembling the Social: An Introduction to Actor-Network-Theory*. Oxford: Oxford University Press.

Lussault, M., and Stock, M. (2010). Doing with Space: Towards a Pragmatics of Space. *Social Geography*, 5(1), 11–19.

Schlögel, K. (2005). *Marjampole oder Europas Wiederkehr aus dem Geist der Städte*. Munich: Hanser.

Schlögel, K. (2009). Die Ameisenhändler vom Bahnhof Zoo:Geschichte im Abseits und vergessene Europäer. *Osteuropa*, 11(2009),53–60.

Stewart, K. (2014). Road Registers. *Cultural Geographies*, 21(4), 549–563.

Urry, J. (2007). *Mobilities*. London: Polity.

11 Drawing the Dardanelles

Art History and Mobilities Studies

Ulrike Boskamp and Annette Kranen

This article develops an analytic view of artistic mobility and its historiography, merging art historical materials and questions with categories from mobilities studies. While on the one hand a critical approach to visual media, particularly historical ones, seems not to be firmly established in the context of mobilities studies, on the other handart historical research is only beginning to explicitly focus on mobility in a historical and critical perspective. The form of mobility that was traditionally considered in European art history was the educational journey to Italy, where artists would study the aesthetic ideal of antiquity and Renaissance art. In the context of globalization discourses, scholars have begun to investigate differently motivated travels as well as other itineraries. While most of these studies concern contemporary artists, recent research also broadened the historical perspective and considered earlier artists' travels. Apart from the classic *Bildungsreise*, diverse motivations for historical artists' mobilities, such as migration, exile, training, or employment, as well as other destinies like Eastern Europe or colonized regions, are being taken into account by a transculturally oriented art history.

Our contribution is situated in this emerging area of research. We refer to Tim Cresswell's 'constellations of mobility' which suggest the investigation of three different aspects: movement, representations, and practices. This conceptual framework forms the basis for a close reading of exemplary historical materials and images. In the context of this volume, our approach presents a historical perspective on the relation between mobilities and visual representations.

Visual media, such as maps or images of landscapes, form a vital part of the multiple and interwoven practices of mobility. Images of travel and places have been given a major share in the complex conceptions of mobility that have emerged from sociological and geographical research in the course of the last decade (Urry, 2008; Cresswell, 2011). In his concept of 'constellations of mobility,' Tim Cresswell distinguishes the three aspects movement, representations, and practices, thereby including media as a constitutive element in the setup of mobility itself. Yet close and critical analyses of visual media in contexts of mobility still seem to be lacking. Cresswell has also argued that "the importance of an historical perspective which mitigates against an overwhelming sense of newness in mobilities research" should be recognized (2011: 163). Still, following the establishment of mobilities studies as a discipline to investigate contemporary phenomena, historical constellations and specificities of mobility have rarely been considered in this context, particularly before the purported epochal threshold marking the beginning of modernity around 1800 (Hasty, 2014). So while on the one hand, an analytical approach to visual representations, especially historical ones, seems not yet established in the context of mobilities studies, on the other handart historical research is only beginning to explicitly take mobilities into account in a historical and critical way (Gludovatz, Noth, & Rees, 2015). A further opening up of the historical disciplines towards mobilities studies and vice versa would help to develop new perspectives in both fields of research. Art history could profit from the conceptual framework of mobilities studies in its current exploration of transcultural approaches. Although, for example, the Rijksmuseum in Amsterdam has developed new displays based on concepts of global exchanges and entanglements, the category of the 'national school' still prevails, structuring parts of academic education and specialization as well as the order of historical art in museums. This conception, based on a container model of national cultures, could be set in motion by consideration of historical movements of people, objects, and ideas. Introducing mobility as a category of analysis enhances the sensitivity for processes of transfer and cultural exchange in historic artistic practices and materials, opening up new perspectives into classic art historical materials and directing interest towards objects that have hitherto escaped these classic categories. And finally the categories and preconceptions of mobility that art history is concerned with could be reconsidered. A prime example is the concept of artists' travels, still dominated by the idea of the educational journey from Northern Europe to Italy for the purpose of studying the art of Classical Antiquity and the Renaissance. This needs to be replaced by a much broader assessment of artists' mobility, which takes into account other destinations, motivations, and conditions of travel.

In order to juxtapose the conceptual framework of mobilities studies with art historical materials and method, our paper presents an experimental setup, confronting two case studies of artists' travels. We will consider these exemplary journeys and their contexts, including the relevant institutions, commissions, topographies, drawings, the distribution of images, and notions of alterity as networked mobilities.

We will focus on representations of the Dardanelles straits originating from two late seventeenth-century French journeys to the Ottoman Empire. In the second half of the seventeenth century, the French had intensified their diplomatic and economic relations with the Ottomans. At the same time, the desire and plan to conquer Ottoman territories arose at the French court of Louis XIV, a plan that was never realized (Omont, 1893; Bilici, 2004). As the crucial passage connecting the Aegean with the Sea of Marmara on the route from Western Europe to Constantinople, the Dardanelles were a focal point of strategic and military interest. Shortly before, during the Cretan

War (1645–1669), it had been the site of siege and sea-battles between the Venetian and the Ottoman fleets. These events contributed to the idea of the Dardanelles as a location of transcultural encounter and military combat between Christians and Muslims during the seventeenth century. Furthermore, as the Dardanelles were the place of ancient Troy and Greek mythology, they were also ascribed a high cultural significance.

These layers of meaning were constructed and sustained through visual representations. Artists traveled from France to the Ottoman Empire with a shared background of ideas, but with diverse motivations and under different conditions. We will focus on visual media representing the Dardanelles with regard to persons on the move (traveling artists), institutional frameworks of their travel (employers, patrons, institutions), and a paradigmatic 'mooring' (the Dardanelles' topography) (Hannam, Sheller, & Urry, 2006).

Two Trips Through the Dardanelles and Their Images: Guillaume Grélot's Artist's Journey (1670–1675) and Étienne Gravier D'Ortières Reconnaissance Mission (1685–1687)

The artist Guillaume Joseph Grélot traveled to Constantinople in 1670 (Grélot, 1680).[1] After a two years' stay in the Ottoman capital, he continued on to Persia, returning back to France in 1675. During his journeys, Grélot drew landscapes, cities, and architectures. While some of his original sketches from Persia have survived in the manuscript of Ambrosio Bembo's travel journal (Bembo, 1676), none are known from his stay in Constantinople. After his return to France, Grélot published an account of his journey to Constantinople that contained elaborate illustrations of the city and its buildings. The book appeared in 1680 and was dedicated to Louis XIV (Grélot, 1680). A large engraving, *Veue de l'Hellespont et de la Propontide* (Figure 11.1), illustrated Grélot's description of the Dardanelles in chapter one.

When the French government decided to prepare for an invasion of the Ottoman Empire, the need for precise information on its topographies, and in particular for maps and images of harbors, coasts, cities, and fortifications, arose. This was the motif for a reconnaissance voyage initiated in 1685.[2] Head of the mission was the officer Étienne Gravier, marquis d'Ortières. The leading draftsman, the military engineer Étienne Plantier, held the responsibility for the production of maps and views intended for the king's war archives. He was assisted by at least one other draftsman. A diplomatic mission dissimulated the true purpose of the trip: Officially, the group's tasks were to accompany the newly appointed ambassador Pierre Girardin to Constantinople (Vogel, 2013) and to inspect the consulates of the 'Échelles du Levant,' Mediterranean port cities where the French merchant communities held special privileges (Grélois, 2001a).[3] In November the group set to sea from Toulon under the command of the captain Du Mené, with a flotilla that included two French war ships.[4] Two prominent French military engineers, the brothers Benjamin and Pierre de Combes, joined them later.[5] For

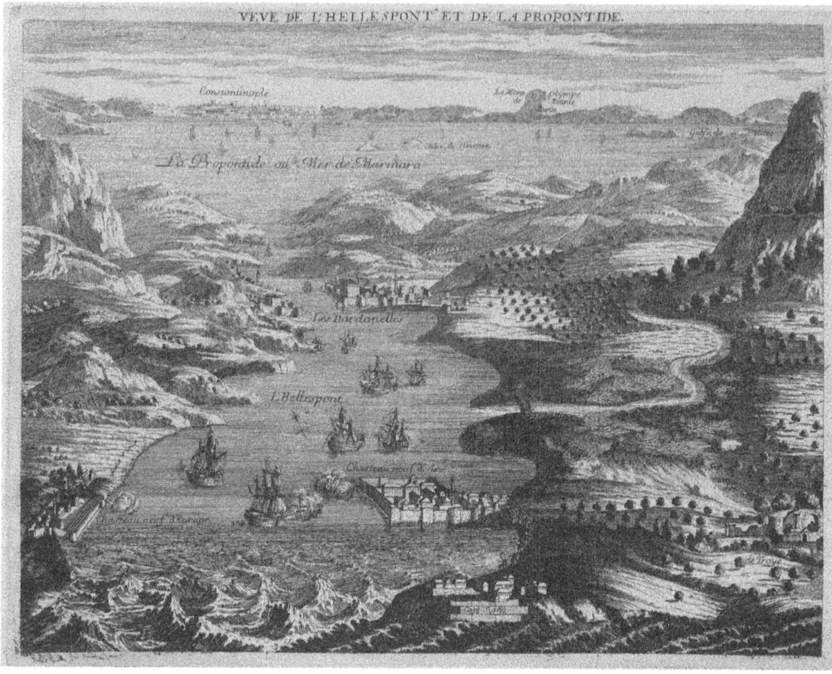

Figure 11.1 Guillaume Joseph Grélot, *Veue de l'Hellespont et de la Propontide*, from Grélot, *Relation nouvelle d'un voyage de Constantinople*, Paris 1680, 41

Source: Hellenic Library, Alexander S. Onassis Public Benefit Foundation, Photo © Aikaterini Laskaridis Foundation.

three days, from 24 to 26 December 1685, they examined the Dardanelles' topography, fortresses, and weaponry. After their arrival in Constantinople on 12 January, the party proceeded with their mission, finally sailing back to France in August 1686 (Bilici, 2004: 144). The trip is extraordinarily well documented. The images and reports from the Dardanelles by Plantier and the brothers de Combes are still kept in French archives (Figure 11.2).[6] Only Gravier D'Ortières' own assessment of the strengths and weaknesses of the Dardanelles' topography and fortresses seems to be missing.[7]

Orders and Dedications—Institutional Frameworks of Travel and Travel Images

The institutional framework is a determining factor not only for the type of mobility, but also for the form and content of the images produced during a journey. In our examples, institutional frames were in the one case necessary to conduct the trip, in the other case to publish the travelogue afterwards. Even though travel images often claim or are believed to be

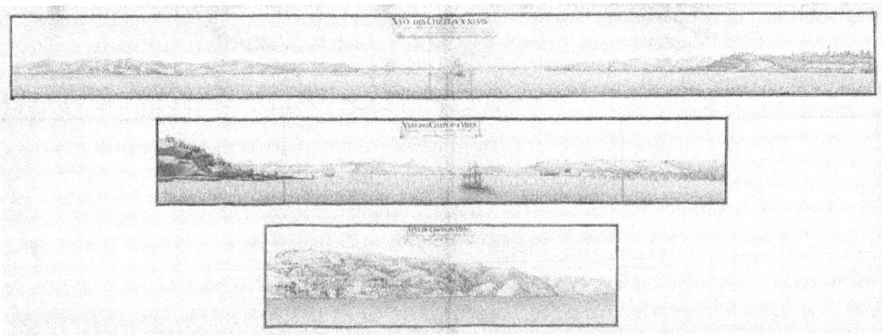

Figure 11.2 Étienne Plantier, Veüe des Châteaux Neufs des Dardanelles en venant de Tenedos; Veüe des Chateaux vieux des Dardanelles, dits Sestos et Abidos, avant de les passer; Veüe du Chateau vieil d'Europe, ou Sestos, prise devant la batterie de la mer, 1686, Paris, Vincennes, Service historique de la Marine

Source: © Service historique de la Défense, DBIB Vincennes SH 109.

authentic representations of places, it is decisive to go beyond the discussion of whether they are reliable or not. Not only the rhetoric of travel and being on site, but also the motives and conditions of mobility, image production, and image publication must be taken into account in order to develop a critical approach to historic travel images.

The journey of our first artist, Guillaume Grélot, seems to have been a private enterprise. He traveled to Constantinople with the French Numismatist Jean Foy-Vaillant, whom he might have accompanied as a servant for some time, but there are no indications that he was under official orders or had institutional funding (Meyer, 2012: 14).[8] After about two years in Constantinople Grélot traveled from Constantinople towards Persia in 1672 in the services of the French merchant Jean Chardin who had employed him to prepare drawings for a travelogue he intended to publish. A controversy ensued between the two men when Chardin refused to pay the artist for his work before the end of their journey. Therefore, Grélot decided to abandon his service after two years without recompense. Stranded in Isfahan without any means, he was lucky to be taken back to Europe by the young Venetian nobleman Ambrosio Bembo (Welch, 2007: 26–32, 360, 361). While no institution supported Grélot's journey, he afterwards dedicated his travelogue to Louis XIV and thereby entered the institutional and ideological framework of French absolutism. Dedicating books to high ranked persons was a common practice in the early modern period and a means for authors and printers to cover their publication costs (Stört, 2006: 85–89; Hiller & Füssel, 2006: 84).

The bird's eye viewpoint of Grélot's engraving of the Dardanelles (Figure 11.1) indicates its origin in mapping rather than in on-site drawing (Gehring, 2014). This might mean that the artist did not produce this image on his trip but rather in the course of the editing of his travelogue in Paris. He could have used earlier printed views of the straits, such as those from reports on the Cretan War that had been distributed by well-known publishers such as Matthaeus Merian (Abelinus & Merian, 1647: Fig. 46).

In contrast, the journey of Étienne Gravier D'Ortières and his draftsman Étienne Plantier took place in the highly institutionalized framework of the French court, whose bureaucracy has not only preserved the resulting images until today, but also the official instructions for the travelers, the reports they sent back, the communications about the journey inside the French marine, etc. The French ministry of war planned the journey and decided about the travel route, the means of transport, and the personnel. Separate 'Instructions' issued by the minister of war assigned their respective tasks to each traveler, stressing the prime significance of gaining exact knowledge about the Dardanelles.[9] The draftsman and engineer Plantier's general commission is clearly defined:

> [His Majesty] wants that in all the places of the Archipelago . . ., he will take plans and maps and will measure exactly the moorings, and carefully examine the entrances of the ports, their distances to the closest dockyards, the transversal of the ports and with which winds one can enter into them.
>
> (Omont, 1893: 198)

Plantier reported the successful visit of the Dardanelles to the minister of war on 10 January 1686 (Omont, 1893: 202–203). Back in Paris, he was ordered to redraw and finalize his sketches from the journey as fast as possible. Thus, for official use, they were copied and colored, captions and compass roses were inserted, and, in the case of the charts of the Dardanelles, the depths of the waters were entered as well as the range and orientation of the Turkish canons. Only these fair drawings, but none of the drafts made on site, are preserved.

The two journeys were conducted and represented under very different conditions. Grélot's case indicates the precariousness of artists' travels in the early modern period, when they were not secured through an official function. Still, the funding and dedication of his printed travelogue might point to an institutional impact on its illustrations. The journey of Gravier d'Ortières and Plantier, on the other hand, was institutionally initiated, highly organized, and extremely well funded.

Imagining Conquest—Representing the Dardanelles as a Military Topography

The Dardanelles, as the only sea passage from Western Europe to the Ottoman capital, were a focal point of imaginations of invasion. After the victory of the European armies against the Turks before Vienna in 1683, the conquest of the Empire appeared attainable. The Muslim Ottomans were perceived as occupants of the Holy Land and the sites of classical antiquity—places that were considered a part of Western European history and identity. This atmosphere of conflict and othering was introduced into the representations of the Dardanelles.

In his engraving (Figure 11.1) Grélot provides an overview of the Dardanelles and the surrounding landscape from an elevated point of view, including the sea of Marmara (here called Propontide) with the city of Constantinople on the horizon. The text underlines the military significance of this topography, presenting the Dardanelles as a 'theatre of war' dating back to antiquity. It refers not only to Homer's Troy, the ruins of which are included in the lower right side of the engraving, but also to the recent naval battles between the Venetian and Ottoman fleets (Grélot, 1680: 14–18, 19).[10] The engraving strikingly exaggerates the dimensions of the four Ottoman fortresses that defend the straits.[11] This mode of enhancing their visibility can be found in many representations of the Dardanelles in the seventeenth century, so the military significance of the straits formed part of its conventional iconography.[12] Emphasized in this way, the fortresses serve as metonymic image representing the well-known Turkish military potency and, at the same time, their vincibility. In his text, Grélot underlines the outdated technology and vulnerability of these architectures (Grélot, 1680: 4–6).

The Dardanelles and Constantinople were usually not represented in one image, not even in older military maps. But Grélot's image includes the city in this landscape of military significance, so that it appears on the horizon as the object of desire and conquest.[13] This iconographic feature is embedded into the context of the volume. In his dedication, Grélot pays homage to the sun king and addresses him as the potential conqueror of Constantinople:

> I hope, that your Majesty who is no less distinguished by his mildness than by the great deeds in the succession of Alexander, will allow me to give him at least the image of the places that he could conquer if he approved to bring there his ever victorious arms.[14]

Grélot's emphasis on the strategic significance of the Dardanelles and his allusion to a possible conquest of the city can be regarded as an extension of this *captatio benevolentiae* to Louis XIV. Additionally, his references to a glorious prehistory of war at the Dardanelles place the French king into a famous genealogy of conquerors.

In contrast to Grélot, the party of Gravier d'Ortières produced maps and views of the Dardanelles straits and of the four fortresses and their weaponry directly on site.[15] Plantier's images from this journey were bound into three volumes that bore the very explicit title *The state of the military places that the muslim princes possess on the coasts of the Mediterranean and of which the maps have been taken by order of the king on occasion of the visit of the Eschelles de Levant which His Majesty has ordered to be undertaken in the years 1685, 1686 and 1687, with the project of invading there and becoming master of it.*[16]

The objective of the mission had been to obtain the highest possible degree of exactness in the representation of the enemy's terrain (Omont, 1893: 203). The iconographies of Plantier's highly finished colored drawings of the Dardanelles strictly follow their military purpose: A map giving an overview of the site and topography is complemented by three views of the straits as visible during the passage of a ship in three different positions (Figure 11.2; Bilici, 2004: 159), and finally, each of the four Dardanelles fortresses is represented on a single sheet, including ground plans and elevations. One separate view is dedicated to the weapons in the new Asian fortress.

The fantasy that perfect images of the enemy's land would enable and even guarantee military success is vividly demonstrated in the separate volume in which Plantier's maps and views of the Dardanelles are kept. It opens with an exquisite large map of the Dardanelles, decorated with anticipations of victory: A cartouche features enchained Turks under a darkened moon symbolizing the Muslim Ottoman Empire, which is overpowered by the sun, i.e. France. The three views of the Dardanelles passage (Figure 11.2) show European war ships heading towards Constantinople. This album, it seems, was destined for the personal use of the sun king himself.

Both examples of topographical views of the Dardanelles straits include the military imaginary of conquest. Grélot's representation of the Dardanelles alludes to a phantasy of invasion, while praising the military potency of Louis XIV. A different but likewise military imaginary is inscribed into Gravier D'Ortières' mission: The belief that the practices of precise representation and measurement will lead to visual media with the potency and the agency to ensure victory.

Traveling Images—Policies of Publicity and Circulation

The meaning, interpretation, and reception of images depend not only on their iconographies, but also on uses and image policies: While Grélot's view of the Dardanelles was part of a commercial enterprise, aimed at a wider public and spread by reproduction through print, Plantier's images from Gravier d'Ortières' mission were intended for a highly restricted access. Furthermore, images can be transferred into new contexts. There they acquire new meanings but also carry indexes of their original circumstances of production and reception, thereby transferring features from one context into another. This happened in both of our cases.

In the case of Grélot, the images and the text of his travelogue were used by other travelers as well as by scholars.[17] His image of the Dardanelles was widely disseminated and copied many times in the decades following its publication in 1680. It appeared in several geographic and historic works, thus transforming into a visual topos in Western Europe. For example, Jacob von Sandrart inserted the view into his history of the Republic of Venice where it served to illustrate an account of the Cretan War (Sandrart, 1686: 120). It was then copied and reused by Dutch publishers for a topographical single leaf print,[18] and also entered into a series of engraved city views, a popular genre in the Netherlands at the time (Peeters, 1686: n.p.). Sixty years later, it reappeared in Jean-Antoine Guer's *Moeurs et usages des Turcs* [. . .], illustrating a description of Constantinople (Guer, 1746–47: 142). Thus Grélot's image of the Dardanelles, published in the context of a travelogue, traveled through various contexts. The frequent reuse of the Grélot image of the Hellespont transformed it into a stereotype, a visual condensation of the struggles of Western European powers with the Ottoman Empire, which carried the inscriptions of military desire from its original publication.

The precise maps and views that were produced following Gravier D'Ortières and Plantier's inspection of the Dardanelles were not only produced secretly, but they were also kept hidden, forming part of a plan for a military aggression against a state with which France sustained close political and economic relationships and exchanges. They must have been stored in the king's archives for later use in case of a realization of the invasion plans—which were given up soon after. Therefore, it is surprising to find Plantier's images of the Dardanelles fortresses reused in a travelogue by the botanist Joseph Pitton de Tournefort, *Relation d'un voyage du Levant*, published in Paris in 1717 (Figure 11.3).

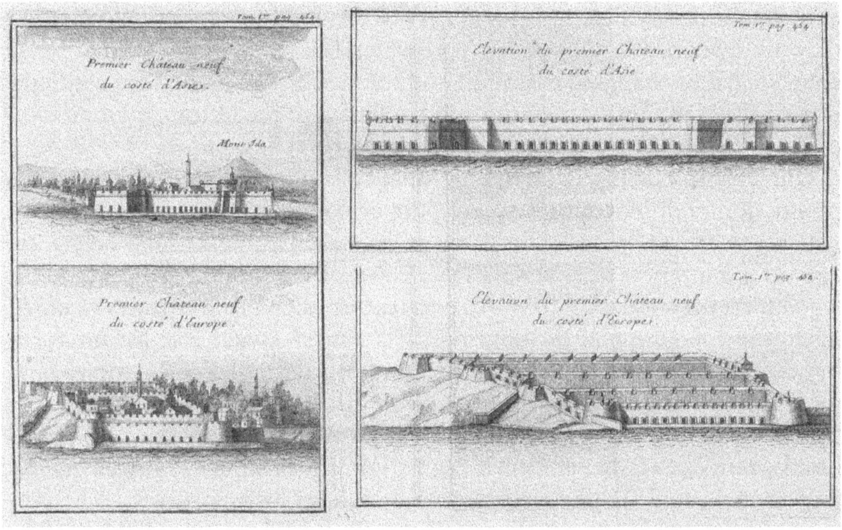

Figure 11.3 Elévation du premier Château neuf du costé d'Asie. Premier château neuf du costé d'Asie. Premier château neuf du costé d'Europe. Elévation du premier château neuf du costé d'Europe, from Joseph Pitton de Tournefort, *Relation d'un voyage au Levant*, Paris: Imprimerie Royale, 1717

Source: Gallica/Bibliothèque Nationale de France.

This author describes the Dardanelles fortresses at length and gives an estimation of their technical backwardness, similar to Grélot's statement thirty years earlier: "Six bombs would be able to demolish these fortresses" (Tournefort, 1717: 162; Grélot, 1680: 4–6). Tournefort had traveled with his own draftsman, Claude Aubriet (Hamonou-Mahieu, 2010: 91–106). To his views, the editors of Tournefort's travelogue added Plantier's plans, produced for military purposes and stored as highly protected and powerful aids for warfare in the king's *Dépot de la guerre*. The new context of the travelogue dissimulated their military origins. Yet the general idea of cultural and religious confrontation and military conflict seems to have persisted through the changes of political circumstances and policies of publication. In spite of the botanical focus of Tournefort's book, this tradition found a condensed expression in the images of the highly fortified Dardanelles, seemingly considered an indispensable ingredient of any report from the Ottoman Empire.

Conclusion

We have described and compared two very different French seventeenth-century artists' travels to the Dardanelles. The images that form part of these 'constellations of mobility' have been examined from three different angles, which were derived both from mobility studies and from art history. Institutional networks have proved vital for the iconography, meaning, and functions of these images in different ways. A military fantasy of invasion and conquest was shown to form part of the representations, even though they are inscribed in very different registers. The images' reuse in different contexts led to changing meanings and references, but also to the continuation of Western European notions of conflict with the Ottoman Empire and of the Dardanelles as a site with military charge.

Our example demonstrates how images from and about travels, their production and dissemination, are part of complex systems of mobilities, which can fruitfully be investigated in historical contexts long before the ages of exploration or tourism. Tim Cresswell's argument from mobilities studies, that representations evoke and reaffirm significant features of movement and topographies or moorings, has played a vital role for interpreting the visual media from the context of journeys of French travelers to the Ottoman Empire. The Dardanelles is constructed and reinforced as a military mooring through a complex system of practices, such as border rituals, flows of traffic, and their regulation, fortification, spying, drawing, representing, and the invocation of their role in historical wars. The conceptual framework of mobilities studies contributes to understanding these processes and to regard travel images (in the broadest sense) as an integral part of them. On the other hand, the art historical analysis of these images in context has demonstrated their importance for an understanding of historic constructions of topographies, including their imaginary charges and visually encoded semantics, thus underlining the potential and necessity of a critical approach to visual media for mobilities studies.

Notes

1. On Grélot's journey most recently: Longino (2015: 108–128); see also: Grélois (2001b) and Meyer (2012).
2. For this journey see Bilici (2004) and Omont (1893).
3. See also catalogue entries 44–64.
4. Du Mené received 'Instructions' dated 15 August which gave him his orders for the trip and presented to him the group he was going to host (Grélois, 2001a: 40). Plantier received his own analogous orders, reprinted in Bilici (2004: 132–133, and 169, note 36). On the background of the French Aspirations in the Levant see also: Haran (2000: 302–307).
5. For the Descombes' position and education see Bilici (2004: 134).
6. Benjamin de Combes, Bibliothèque Nationale de France, département cartes et plans: Carte de l'Hellespon ou canal des Dardanelles, distant de 200000 mil ou 67 lieux de Constantinople avec ses quatres châteaux et les sondes, 63,5x 92 cm (GE SH 18 PF 98 DIV 3 P 1); [Carte de l'Hellespont], 19 x 30,6 cm (GE SH 18 PF 98 DIV 3 P 1/1 D); Château neuf d'Europe avec partie de son bourg, 63,5 x 91,5 cm (GE SH 18 PF 98 DIV 4 P); Plan du chateau neuf d'Asie . . ., 63,5 x 91,5 cm (GE SH 18 PF 98 DIV 6 P 1); Plan du vieus chateau d'Europe, 63,5 x 91,5 cm (GE SH 18 PF 98 DIV 5 P 1); Plan du Vieux Chateau d'Asie Abidos, 63,5 x 91,5 cm (GE SH 18 PF 98 DIV 7 P), Plan de la ville et cha[tea]u de Galipoly avec son arsenal de marine, 56,2 x 41 cm (GE SH 18 PF 98 BIS DIV 2 P 1 D).
7. He refers to this report in his description of those on the Bosporus, but the document seems to be lost, see Bilici (2004: 269): "J'ay parlé suffisamment du fort et du faible des Châteaux des Dardanelles et de la ville de Constantinople et de la facilité qu'il a de brusler, il faut maintenant expliquer les forces nécessaires pour l'exécution de ce dessein."
8. Grélot mentions Vaillant only a few times in his travelogue without giving any information on their relation, see Grélot (1680: 39, 44, 59–61, 223).
9. Omont (1893: 196–198) reprints a large part of the instructions for the captain Du Méné and for the draftsman Plantier. In Plantier's instructions we read:

    > Il doit principallement s'appliquer à examiner le passage des Dardanelles, la force des Châteaux qui y sont, et observer quels vents doivent régner pour y passer, à quelle distance l'on en est éloigné en passant, et s'il est facile de tenter ce passage avec plusieur vaisseaux et galères, quelles conjonctures de temps il faudroit prendre pour cela, s'il y a des batteries considérables qui regardent la mer, quel nombre de pièsces de canon il y a dans les batteries et quelles garnisons il y a dans les Châteaux.

10. On the early modern term 'war theatre' see Füssel (2008).
11. Fortresses functioned as markers of possession and governance in early modern representations of the Ottoman Empire, see Brummett (2015: 128–150).
12. For example: Breuning von Buchenbach (1612: n.p.); Sandys (1615: 24); Beauvau (1624: 33); Deshayes (1624: 300).
13. On the connotation of cities, especially Constantinople, as female see Longino (2013: 125).
14. Grélot (1680: Dédication, n.p.):

    > j'espere que VÔTRE MAJESTÉ qui n'est pas moins élevée par sa douceur que par ses grandes actions au dessus d'Alexandre, me permettra de luy donner au moins la figure des lieux qu'elle pourra soûmettre en effet lorsqu'elle trouvera bon d'y porter ses Armes toûjours victorieuses.

15. For the eminent meaning that the employment of the brothers de Combes constituted, see Bilici (2004: 134–135). They gave a separate proposition to the

minister of war (BNF, ms. fr. 5580 'Mémoire concernant le détroit de l'Elespont [. . .]' of April 15, 1686), see Bilici (2004: 163).

16. *Estat des places que les princes Mahométans possèdent sur les costes de la mer Méditerranée et dont les plans on esté levez par ordre du Roi à la faveur de la visite des Eschelles de Levant, que Sa Majesté a fait faire les années 1685, 1686 et 1687, avec les projets pour y faire descente et s'en rendre Maistre*. The full text is printed in Bilici (2004: 188–310). Of this report, not all of the highly finished maps and drawings were completed. Only the first volume and maps are known, and it was still in the course of being written in August 1688, see Grélois (2001a: 41). Further images by the brothers de Combes and a very lavish volume were also integrated into the king's archive.
17. For example the Dutch traveler Cornelis de Bruijn repeatedly refers to Grélot's book in his own travel account, see Bruijn (1698: e.g. voorreeden, n.p.).
18. Single leaf print by Gaspar Bouttats, *Het ghesicht van den Hellesponten van de propontide*, c. 1684–1692, 14 x 26.9 cm, London, British Museum. Also see Ziegler (2014).

References

Abelinus, J. P., and Merian, M. (1647). *Theatrum Europaeum*, 5, Frankfurt: Merian.

Beauvau, H. de (1624). *Relation Iournalière du Voyage du Levant* [. . .]. Nancy: Jacob Garnich.

Bembo, A. (1676). *Viaggio e giornale per parte dell' Asia di quattro anni incirca fatto.* Manuscript preserved in the James Ford Bell Library, University of Minnesota.

Bilici, F. (2004). *Louis XIV et son projet de conquête d'Istanbul.* Ankara: Türk Tari Kurumu.

Breuning von Buchenbach, H. J. (1612). *Orientalische Reyß Deß Edlen unnd Besten Hanß Jacob Breüning von und zu Buochenbach* [. . .]. Strasbourg: Johann Carolo.

Bruijn, C. de (1698). *Reizen van Cornelis de Bruyn door de vermaardste deelen van Klein Asia, de eylanden Scio, Rhodus, Cyprus, Metelino, Stanchio, etc.* [. . .]. Delft: Henrik van Krooneveld.

Brummett, P. (2015). *Mapping the Ottomans: Sovereignty, Territory, and Identity in the Early Modern Mediterranean.* New York: Cambridge University Press.

Cresswell, T. (2011). Towards a Politics of Mobility. In M. Hvattum and J. Kampevold (Eds.), *Routes, Roads, and Landscapes* (pp. 163–177). Farnham: Ashgate.

Deshayes, L. (1624). *Voiage de Levant fait par le commandement du Roy en l'année 1621 par le Sr. D.C.* Paris: Adrian Taupinart.

Füssel, M. (2008). Theatrum Belli: Der Krieg als Inszenierung und Wissensschauplatz im 17. und 18. Jahrhundert. *Metaphorik*, 14, 205–230.

Gehring, U. (2014). Painted Topographies: A Transdisciplinary Approach to Science and Technology in Seventeenth-century Landscape Painting. In U. Gehring and P. Weibel (Eds.), *Mapping Spaces: Networks of Knowledge in 17th Century Landscape Painting* (pp. 22–101). München: Hirmer.

Gludovatz, K., Noth, J., and Rees, J. (Eds.) (2015). *The Itineraries of Art: Topographies of Artistic Mobility in Europe and Asia 1500–1900.* Berlin: Fink.

Grélois, J. P. (2001a). En Orient, l'offensive diplomatique. In M. F. Auzépy and J. P. Grélois (Eds.), *Byzance Retrouvée. Erudits et Voyageurs français (XVe–XVIIIe siècles)*, (pp. 39–43). Exposition à la Chapelle de la Sorbonne, Paris, 13 août–2 septembre 2001.

Grélois, J. P. (2001b). Un dessinateur Français en Orient, Guillaume-Joseph Grélot. In M. F. Auzépy and J. P. Grélois (Eds.), *Byzance Retrouvée: Erudits et Voyageurs français (XVe–XVIIIe siècles)* (pp. 46–48). Exposition à la Chapelle de la Sorbonne, Paris, 13 août–2 septembre 2001.

Grélot, G. J. (1680). *Relation nouvelle d'un voyage de Constantinople: Enrichie de plans levés par l'auteur sur les lieux, et des figures de tout ce qu'il y a de plus remarquable dans cette ville: Présentée au Roy par Guillaume Joseph Grélot.* Paris: P. Rocolet.

Guer, J.-A. (1746–47). *Moeurs et usages des Turcs, leur religion, leur gouvernement civil, militaire et politique: avec un abrégé de l'histoire Ottomane.* 2nd vol. Paris: Coustelier. Available at: http://hdl.handle.net/2027/nyp.33433000098925.

Hamonou-Mahieu, A. (2010). *Claude Aubriet, artiste naturaliste des Lumières.* Paris: CTHS Sciences.

Hannam, K., Sheller, M., and Urry, J. (2006). Editorial: Mobilities, Immobilities and Moorings. *Mobilities*, 1(1), 1–22.

Haran, A. (2000). *Le lys et le globe: Messianisme dynastique et rêve imperial en France aux XVIe et XVIIe siècles.* Seyssel: Champ Vallon.

Hasty, W. (2014). Metamorphosis Afloat: Pirate Ships, Politics and Process, c. 1680–1730. *Mobilities*, 9(3), 350–368.

Hiller, H., and Füssel, S. (Eds.) (2006). *Wörterbuch des Buches.* 7th edition. Frankfurt: Vittorio Klostermann.

Longino, M. (2013). Constantinople: The Telling and the Taking. *L'Esprit Créateur*, 53(4), 124–138.

Longino, M. (2015). *French Travel Writing in the Ottoman Empire: Marseille to Constantinople, 1650–1700.* New York and London: Routledge.

Meyer, G. (2012). Un voyage au Levant effectué par ordre de Louis XIV: Jean Foy-Vaillant dans l'Empire ottoman. *Synergies Turquie*, 5, 13–26.

Omont, H. (1893). Projets de prise de Constantinople et de fondation d'un empire français d'orient sous Louis XIV. *Revue d'histoire diplomatique*, 7, 195–246.

Peeters, J. (1686). *Het ghesicht van den Hellespont en van de Propontide, Korte beschryvinghe, ende aen-wysinghe der plaetsen in desen boeck, met hunnen teghenwoordighen Standt, pertinentelijck uyt-ghebeldt, in Oostenryck.* Antwerp: J. Peeters.

Sandrart, J. (1686). *Kurtze Beschreibung Von dem Ursprung, Aufnehmen, Gebiete, und Regierung der Weltberühmten Republick Venedig* [. . .]. Nürnberg: Sandrart. Available at: www.mdz-nbn-resolving.de/urn/resolver.pl?urn=urn:nbn:de:bvb:12-bsb10804650-4.

Sandys, G. (1615). *Relation of a Iourney Begun An: Dom: 1610. Foure Bookes: Containing a Description of the Turkish Empire, of Aegypt, of the Holy Land, of the Remote Parts of Italy, and Ilands Adioyning.* London: W. Barrett.

Stört, D. (2006). Form- und Funktionswandel der Widmung: Zur historischen Entwicklung und Typologisierung eines Paratextes. In V. Kaurkoreit, M. Atze and M. Hansel (Eds.), *"Aus meiner Hand dies Buch* [. . .]*": Zum Phänomen der Widmung* (pp. 79–112). Wien: Turia + Kant.

Tournefort, J. P. de (1717). *Relation d'un voyage du Levant.* 2nd vol. Paris: Imp. Royale.

Urry, J. (2008). Moving on the Mobility Turn. In W. Canzler, V. Kaufmann and S. Kesselring (Eds.), *Tracing Mobilities: Towards a Cosmopolitan Perspective* (pp. 13–23). Aldershot: Ashgate.

Vogel, C. (2013). Gut ankommen: Der Amtsantritt eines französischen Botschafters im Osmanischen Reich im späten 17. Jahrhundert. *Historische Anthropologie*, 21(2), 158–178.

Welch, A. (Ed.) (2007). *The Travels and Journal of Ambrosio Bembo*. Berkeley, Los Angeles and London: University of California Press.

Ziegler, P. (2014). A New Genre of Cartographic Books: The Atlas Production in the Low Countries. In U. Gehring and P. Weibel (Eds.), *Mapping Spaces: Networks of Knowledge in 17th Century Landscape Painting* (pp. 145–155). München: Hirmer.

12 Ghosts of Our Consumption

The Debris Project

Lee Lee

Initiated by Lee Lee and maintained in collaboration with arts and environmental organizations, the Debris Project focuses on learning about and representing the health and environmental impacts of plastic pollution. Over the course of five years the project has been building collaborations between communities across the Pacific, Atlantic, Caribbean, Mediterranean, and North seas, where people have been engaging in the concerns related to the circulation of plastic debris. In short, the practice combines cleaning plastic off coastlines and rivers, educating children and adults about the global circulation of plastic, and making collaborative artworks of oceanic life using the collected plastic. On a network scale the artistic practice is a traveling one. Where the results of diverse local processes are gathered to make an ongoing visual archive of both artworks and images, it shows emerging global commitments to envision possible futures for urban patterns around the use of plastic. The design of the project itself follows the material flows of plastic around the world and deep into the communities who live near the shores where debris is washed up. This chapter is a summary of some of the methodological inspirations, experiences, and reflections that make up the Debris Project.

It was at the Cosmobilities conference in Copenhagen that I first met John Urry. I was feeling like an outsider, because I'm a practitioner not an academic and because my practice stems from a deep ecological concern instead of being arrived at through the field of mobilities studies. However, our warm conversation reaffirmed the relevance of the Debris Project as part of the discourse around mobilities. We were talking about how Mr. Urry explored futures, ones he imagined based on different equations being realized in the present. I shared with him that one of the foundations of the Debris Project was the chemical future expressed in *Our Stolen Future* (Colborn, Dumanoski, & Peterson, 1996). Looking down and shaking his head in a way that conveyed sorrow, Mr. Urry expressed familiarity with the work and said he needed to delve deeper into the ideas therein.

Plastic Pollution

It is a terrifying future when we consider the emerging evidence of severe biological disruptions being realized because of petrochemicals. In addition to the chemical cocktails that are added to plastic during the manufacturing process, chemicals that have been released into watersheds and ultimately the ocean are absorbed by plastic marine debris, with some concentrations reaching millions of times the concentration of chemicals than in the ambient waters surrounding it. Founder of the Algalita Marine Research Institute, Captain Charles Moore, wrote about this in his book, *Plastic Ocean* (Moore & Phillips, 2011) after his discovery of the Great Pacific Garbage Patch. His discovery led to extensive studies of the far-reaching impacts of a mobilities-centered culture. Swept up in global weather patterns and ocean currents, the material and chemical impact of plastic pollution may be recognized as the materialized ghosts of our consumption that haunt the most remote areas of the planet. Expressing extreme mobility, Persistent Organic Pollutants (POPs) found in petrochemicals travel vast distances on air or water currents, as well as animal migration. When chemical-laden plastic is accidently consumed by wildlife, the POPs de-sorb into their flesh and are biomagnified (increasing in concentration) as they move through the marine food web (Rochman, 2015). In the same way that chemicals are absorbed by plastic, which is essentially a synthetic fat, they can easily attach to the fat of animals. Thus, chemical bioaccumulation becomes more concentrated in animals who carry a lot of fat. Because of this, Arctic wildlife bear the brunt of chemical exposure despite their remote habitat. In a similar way, the highly mobile chemicals released anywhere in the world may also arrive on our own doorsteps. The issue of plastic pollution is quite literally a global network of organic, climactic, and chemical mobilities. Plastic pollution infuses environments and bodies, staying there for generations or centuries, even though they are initiated by only a few generations of modern mobilities-centered lives.

From the start, plastic as a physical material is intimately linked to petrochemicals. In *Societies Beyond Oil* (Urry, 2013), Mr. Urry makes the argument that contemporary society is entirely built from the consumption of oil. He describes how oil is crucial for at least 95% of food production and distribution, including the chemical ingredients that make up pesticides and fertilizers as well as moving the food to market. Oil is also an ingredient of plastic used in 95% of manufactured goods and packaging, as it can be transformed into an extraordinary range of materials. This link exasperates both problems of chemical pollution and the physical waste caused by an inordinate use of single-use plastic.

It is this ubiquitous materialism of plastic pollution that I have been using to create an artistic practice since my son was born eight years ago. My concern over the overuse of chemicals broadened to plastic pollution as I came to understand the immensity of physical debris that clogs waterways and entangles wildlife. We gain a better understanding of mobilities through deconstructing its essential materials. Mimi Sheller demonstrates this in *Aluminum Dreams* (2014) by offering an in-depth exploration of the material contributions of aluminum to mobilities, as well as the social and environmental impacts of the production. Like aluminum, plastic makes

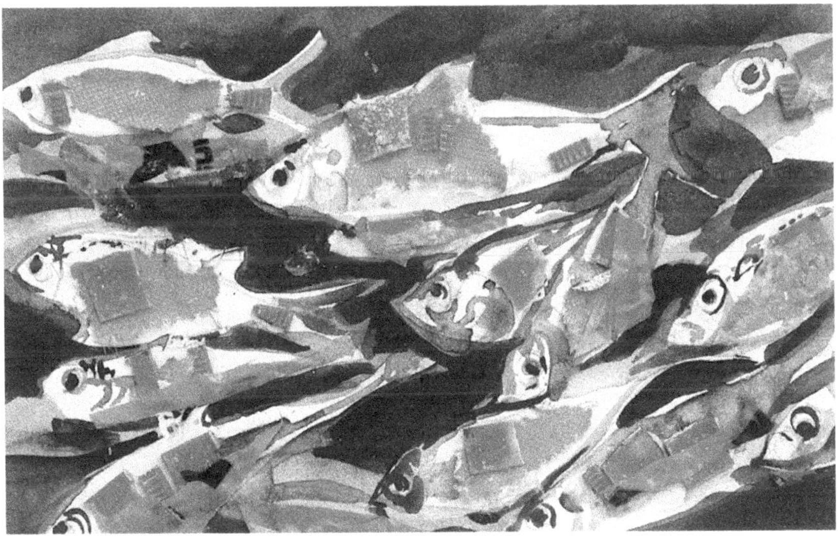

Figure 12.1 The Debris Project, Cosmobilities Installation, Aalborg University. This image of alewives, painted by Lee Lee, represented the Debris Project in an exhibition on Biodiversity and Extinction hosted by Art Science Collaborations at the New York Hall of Science

Source: Lee Lee.

mobilities possible. From the casings to our gadgets, to the machines that move us and the products we consume, plastic fills our mobilities-centered lives. It has ecological value because we don't have the natural resources to support our population without it. In fact, a strong argument may be made that it is too valuable to be discarded after only a single use.

Methodology

The Debris Project was structured as an open sourced creative platform to explore the environmental impacts of plastic pollution. It looks at the residues left by a convenience-oriented consumer culture that is central to a fast-paced urban lifestyle, and challenges misplaced cultural notions of 'waste' and 'disposability.' In developing the methodology for the project, I looked to the Slow Food Movement (2016). The effectiveness of this global network consists of work accomplished by local practitioners. The leaders of the movement contextualize the neo-colonial power structures that threatens food security, while helping local communities to develop their own frameworks that effectively address local issues which stem from an industrialized agriculture (Petrini, 2015). The Debris Project is not imposing a specific predefined solution. There is no 'model' that is offered for engagement. The central idea is to use plastic debris to maintain locally specific cultural representations of marine life across a wide range of cultures. Like Slow Food, the Debris Project is designed to work through collaborating with organizations who have a demonstrated long-term commitment to addressing plastic pollution in their communities. Representations that are gathered from these collaborations contextualize the global scope of the issue and are pieced together in exhibitions that engage audiences on an international level. Through such simple and concrete collaborative efforts, new relationships could potentially arise locally and across participating locations.

As an example of a strong local effort to address plastic pollution, the Voluntary Artists Studio of Thimphu is developing ways of talking publicly about trash that challenges people to see how their relationship with waste is transforming. It was the children from the community who initiated the concern, and the collective responded by developing public installations, student workshops, and the implementation of a recycling program that promotes a micro-economy for elders. Along the Raidak River, they transformed a former dump into a sculpture park, where the sculptures are made from reclaimed material found on site. This kind of in-depth, multifaceted effort inspires the methodological approach of the Debris Project. I had the opportunity for creative exchange during a workshop I led with the youth involved with the collective. While their community approach in addressing plastic pollution is strong, they were eager to learn how to insert their actions into the larger sphere of global creative exchange.

The Endocrine Disruption Exchange (2016) founded by Theo Colborn also informs the methodology of collaboration. Dr. Colborn was the first to recognize the trans-generational effects of endocrine disruption, meaning that when I build up a chemical body burden, it is my children and grandchildren that bear the brunt of the impacts. Our health is at stake, and the effects are only beginning to emerge. Since it takes so long to study species across generations, the Endocrine Disruption Exchange was set up as a platform for scientists to share findings in order to support evidence from other endocrinologists in a rare example of open sourced science. Similarly the Debris Project has no expiration date or targeted age groups, but includes work from school children, activists, researchers, and pensioners.

Collaborative Solution Building

The Debris Project was originally conceived during a residency at the Chateau de La Napoule in France. La Napoule Art Foundation asked resident artists to create interactive works geared towards youth. This collaboration is growing into a gathering of concerned voices from all over the world, around the universal exposure to plastic. Through working outside traditional boundaries of disciplinary practices, the Debris Project enacts a step beyond activist rhetoric, while remaining critical to the subject matter. Through this project I focus on building creative engagements that are cultivated by artists,

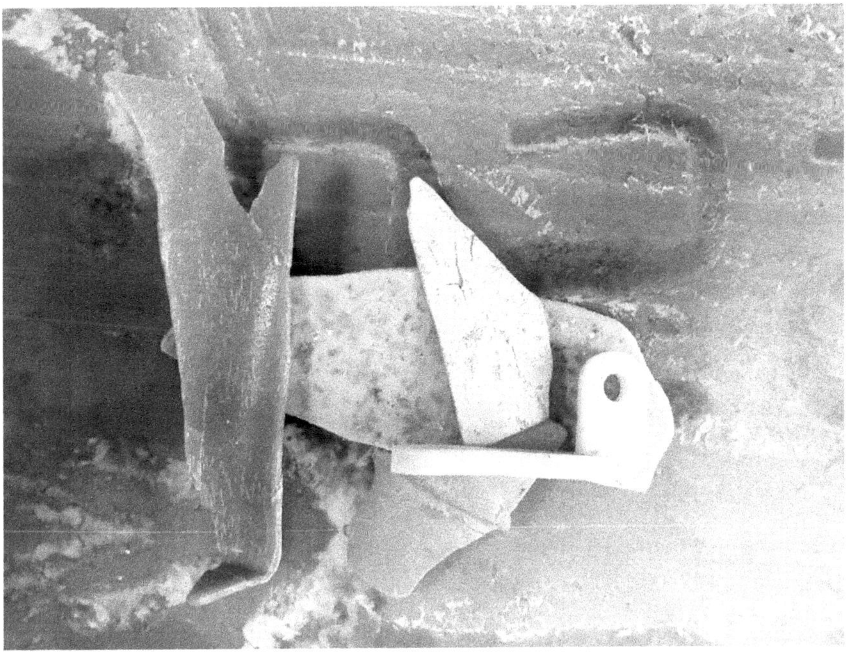

Figure 12.2 Marine debris collected on a Sharkastics beach clean in Maui was used to make a fish by a student from the Ricks Center for Gifted Children in Denver. The work was photographed on a background image of a buoy that was washed ashore in Maine, broadening the geographic representation within the work, reflecting the scale of the issue

Source: Lee Lee.

educators, and scientists. The artistic process starts from the simple social situation of a local beach cleanup. Material gathered during the cleanup is used by teachers in school classrooms or facilitated gatherings in other venues to produce physical artworks and installations. The practice consists of a local process of cleaning, teaching, and re-assembling into art and has been carried out on several continents. There is a physical installation made up of images documenting these processes spanning the globe, as well as an online platform for sharing artistic innovation and scientific evidence across locations and communities.

As a hands-on way for young people to learn about the impact of plastic, they are encouraged to contribute to solution building through shifting habits and innovating new cultural paradigms. By integrating ideas developed by activists like Beth Terry, who shares her ideas and extensive resources in her book *Plastic-Free* (Terry, 2012) and on her blog (Terry, 2016), participants arrive at solutions by learning to adapt their daily decisions around

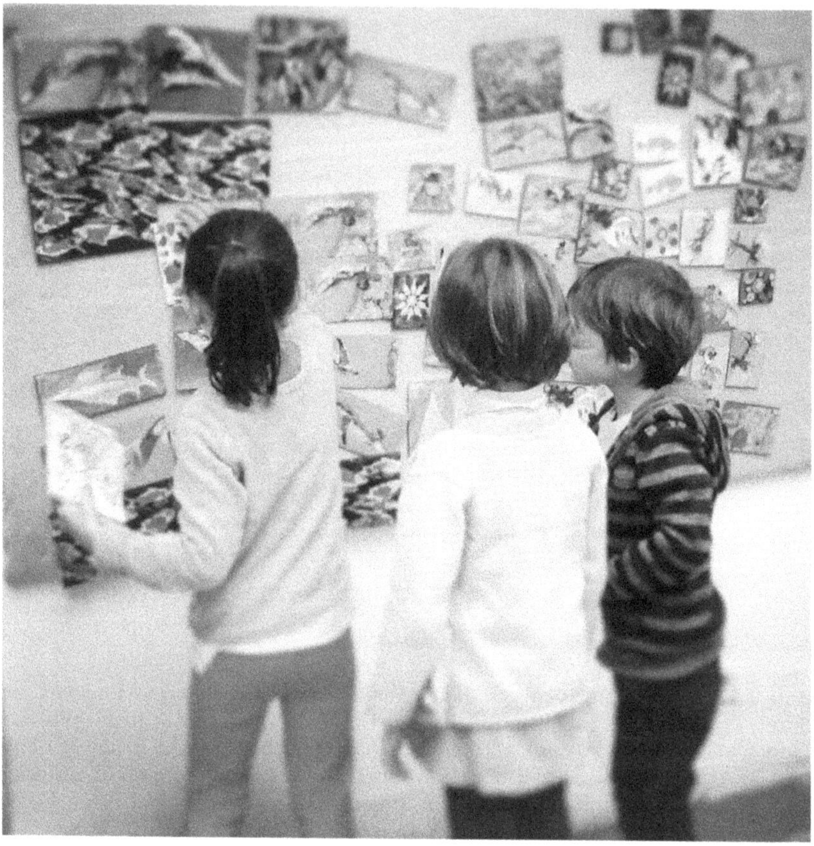

Figure 12.3 Children exploring the marine food web through an interactive installation at the Chateau de la Napoule, France

Source: La Napoule Art Foundation.

consuming plastic by exploring alternatives to the convenience-oriented culture that produces excessive amounts of waste. Through becoming aware of the impact on ecologies, cultures, and our health, they may realize the interconnectedness of ecologies, organisms, and the material impacts imparted by plastic. As Susan Freinkel describes in *Plastic: A Toxic Love Story* (2011), only in the last couple of generations have we shifted towards a 'throw away' culture, so it is possible to reframe cultural attitudes towards valuing materials made from limited resources within a similar timespan.

Education

In order to address misplaced notions of disposability, it is necessary to encourage a reconsideration toward waste at an early age. Children are essential partners in environmental activism, as they carry the issues to the hearts of their families. They also demonstrate a high degree of concern both for wildlife and their own futures. Edward Humes describes trash as the biggest legacy we are leaving for our children in his book *Garbology* (Humes, 2013). My experience in with working with kids is that once they become aware of this legacy, they don't like it. It takes time and repetition to reach a young audience with solutions to such complex problems. So it is important to collaborate with organizations that maintain a long-term, on-the-ground practice, and regularly engage with students. Integrating the Debris Project into their programming transforms the practice into a hands-on tool of expression to reinforce ecological curricula. In recognition that children learn better when engaged in an activity than passively receiving information, the project invites the students to actively participate in a larger response to plastic pollution. The California-based organization Save Our Shores has integrated the Debris Project into their ongoing work of organizing weekly beach cleans, orchestrating classrooms engagements, and participating in policy work to confront commercial pollution sources. The creative works created through their public engagements were installed as an exhibition at the National Oceanic and Atmospheric Administration's (NOAA) Monterrey Bay National Sanctuary Exploration Center in Santa Cruz. The installation allowed local students, community organizers, and the people who regularly participated in the weekly beach cleans to situate their own work within an international context and gain a sense of the global scope of plastic circulation. Inviting the public to learn more about the issue in this free educational center helps realize the NOAA institutional mission to educate the public on issues confronting the marine sanctuary.

The conservation science community has recognized a need for stronger tools of social engagement to help communicate their findings to a broader public. The Society of Conservation Biologists is presenting a visual representation of the Debris Project during their International Marine Conservation Congress in Newfoundland. They express their direct intention that "to conserve the world's oceans we must go beyond science, and use it to inform policy and management, and ultimately to catalyze change" (2016). The Association for Environmental Studies and Sciences (AESS) was also proactive in engaging creative dissemination methods during their 2014 conference, 'Welcome to the Anthropocene: From Global Challenge to Planetary Stewardship,' where they promoted artistic interventions and wove socially engaged, science-based artist presentations into their panels. The Debris Project was presented as an unconventional way to educate the public about plastic pollution through a collaboration that echoed the range of the material issue.

Collaborating with these kinds of proactive educational organizations has led to the development of educational 'toolkits.' The toolkits consist of actual plastic pollution gathered from remote shores, printed textures of larger pieces of plastic pollution, and images from the existing installation that may be used for collage. To gather the hard plastic debris, we participate in beach cleans, like the ones regularly hosted by

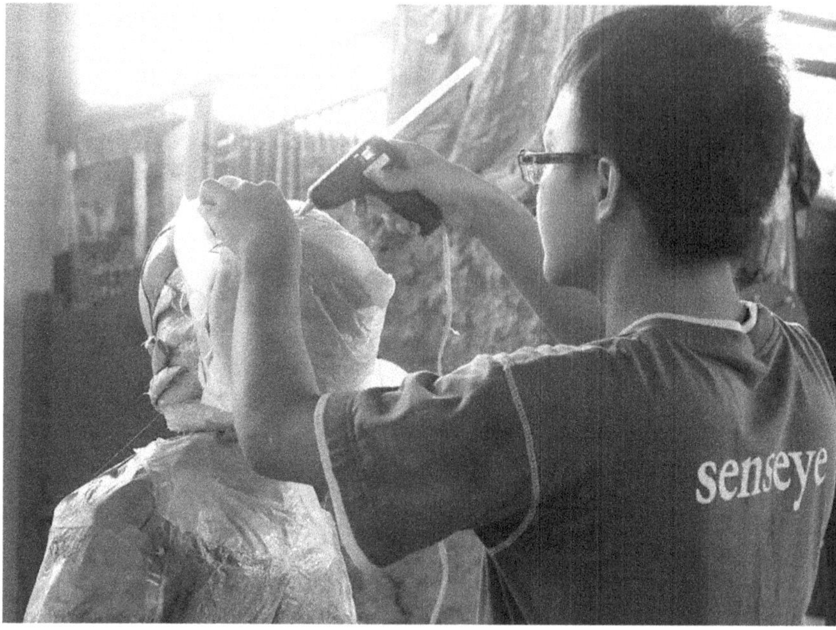

Figure 12.4 A student from the Stratton ABC Foundation uses repurposed plastic waste to construct the 'Plastic Demon' that migrated around Northern Thailand to raise awareness of plastic pollution

Source: John Cope, Stratton ABC.

Sharkastics on Maui where we collected debris from both sides of the Pacific that washes ashore en masse on Wailuku Beach, which catches the ocean current and the open ocean debris carried within. I work with the organization to develop activities that are relevant for their community and often our initial ideas bloom into inspired acts of social engagement. During a Surface Arts Residency at the Rumpueng Community Artspace in Chiang Mai, local workshops with a myriad of organizations inspired the continuation of programming that integrated the Debris Project in a way that promoted public engagement around the issue through outreach by youth participants. Not only did the students from the Stratton ABC home for at-risk youth have the chance to contextualize their own work through a pair of Debris installations, they took it a step forward by creating a 'Plastic Demon' sculpture named Chevy that is currently migrating throughout Northern Thailand to raise awareness of the issue on the streets. Adapting the Debris Project to educational purposes has transformed its methods into a variety of examples of cleaning up plastic pollution, repurposing the material into creative works, and displaying the work creatively in public.

Integration

In exploring the use and misuse of the material by a wide range of cultural appropriations, it becomes evident that everyone has something to offer to building solutions, just as there is room for improvement in each sector. Instead of pointing fingers in blame towards the 'other,' participation is radically inclusive. In a sort of role reversal, those who dwell in 'backwaters' of globalized culture through the Global South are encouraged to be teachers in resourcefulness and best use of materials. Through encouraging collaboration which maintains geographic specificity, the project reflects the scope of the problem by integrating culturally specific representations from a wide range of places. For example, there is no other place than Haiti that could offer an authentic representation of the cultural icon Vodoun Lwa, Boussou, protector of the seas. Beyond the integration of cultural iconography, material specificity is also encouraged. When in Southeast Asia, the primary material integrated into workshops was fused plastic bags because that was the most overused and wasted material. The strength of this part of the Debris Project lay in gathering a wide range of cultural 'voices' around this singular material that has become pervasive in the waters that connect us.

The issues of plastic pollution are driven by urban consumption while the impacts are often remote. As a recognition of this connection between urban and rural spheres, the Debris Project is installed in rural areas that are in the vicinity of debris 'hotspots.' Debris was installed during the spring of 2016 at the Gallery Route One, which is situated in Point Reyes Station near the National Seashore. Kehoe beach, on the National Seashore, is located in such a way that it catches Pacific currents and the open ocean marine debris that washes ashore. Because of the proximity, the gallery has had a long-term dedication to promoting shoreline ecological issues and actively engages local students and tourists to reconsider their relationship with trash.

Inspiring Action: Message in a Bottle

Storytelling is an essential component to inspire action. During a panel hosted by the Cosmobilities Network on Artistic Intervention, the sound-artist and researcher Samuel Thulin defined this kind of approach as 'ecology of narrative,' where we send out seedstock to allow stories to proliferate. Debris stories are presented on the website (www.virtualvoices.org), and artworks and sketches are shown on Instagram (@the_debris_project). While this content reflects the issues, there is a complexity involved in building solutions or stimulating cultural changes, which involves working with the Debris Project as a part of a larger ecology of narratives.

The Debris Project has shown that the mobilities of a story inspire action. While the project has been active, a number of people participating in the project's processes have continued the efforts in their own practices. Some of the creative ideas that grow from re-appropriations within creative practice spark a theme like a 'message in a bottle' which is picked up and transformed in the hands of new collaborators. Native to Tortola, the multimedia artist and historian of Caribbean culture Aragorn Dick Read maintains a beachfront workshop on Trellis Bay in the British Virgin Islands. He was inspired by the Debris Project to execute a performance comprised of a 'message in a bottle' from the sea during one of his New Year gatherings. A large crowd gathered in anticipation of bonfires being lit within his large-scale iron sculptures, installed in the shallow waters off his beach. As the crowd peered across the dark waters, hundreds of glowing PET bottles washed ashore at their feet just before the sculptures burst into flame. This performance inspired Denver Public Schools arts educator Kristin Heeres to develop a variation with her students at Holm Elementary, which serves a largely marginalized, immigrant population in Denver. The class made representations of marine life and placed them in PET bottles, which were then installed at the Denver Aquarium. They were invited to see their work and take a tour, which was for many the first direct exposure to live marine animals. Most students were left in awe. In order to cultivate a concern for the environment, it's vital to feel inspired by the natural world, and the Debris Project offered that opportunity in this case. The 'message in a bottle' theme was replicated in Santa Cruz by the organization Save Our Shores during a youth workshop. The high school students studied the issues and composed written messages to place in PET bottles, helping to cement their understanding of the issues and facilitate an expression of their concerns. Migrating across the Pacific, Art Relief International integrated the theme into their curriculum and brought it into schools across the city of Chiang Mai, Thailand. Their work culminated into an installation at Thapae East which took the form of a giant bottle. Tiles created during their workshop were integrated with tiles from the broader project and were pieced together as a loose skin on the bottle. The sculpture was then filled with plastic salvaged from the wastestream out of their own office as an admirable self-reflection on the amount of waste generated from within. The project inspired Art Relief's Creative Arts Director, Emma Gabriel, to integrate repurposed material over the long-term in order to continue reinforcing awareness of material with her students. To know the effects of a creative practice it is important to understand that stories and knowledge travel, and that the effects of those travels are hard to prescribe. In the case of the Debris Project a pattern of engagement can be traced in multiple directions and types.

Installation

The visual installation of the Debris Project is meant to be light, flexible, and easy to display. Digital images are gathered from around the world and transferred onto repurposed foamcore tiles that make up the foundation of the installation. The scale of the tiles are small tiles to reflect the nature of the way plastic photodegrades, or breaks down into tiny confetti-sized pieces that are spread through ocean gyres. This allows for mobility of the physical installation as it may be broken down into a small box that can be packed in a suitcase or shipped without much expense. In order to keep the participation in the project accessible at a distance, participants from around the world can e-mail digital files of images they'd like to include in the installations. This approach allows makers to keep original works intact, and as a conceptual element to the installation, the physical work is literally dispersed across the globe. When 'released into the ocean' of traveling exhibitions the images become open to transformation in the spirit of collaboration. Echoing the processing of material in the actual ocean, the pieces are copied, distressed, reworked, torn, and mended by participants.

Even as the Debris Project exists on a platform of technological and mechanical reproduction, the work maintains a handmade quality through an imperfect transfer process. The creatures created throughout the project attain a rubbed-out look so that their distressed appearance echoes the corporeal impacts of plastic. Without relying on the sensational but disturbing images of entangled wildlife, this part of the process offers a representation

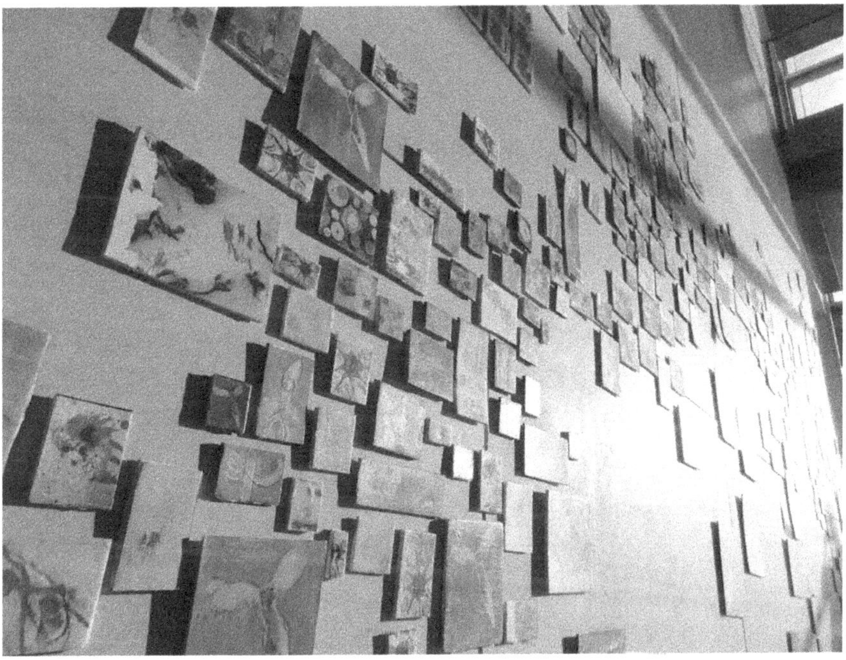

Figure 12.5 Installation of DEBRIS at the Cosmobilities conference at the Aalborg University in Copenhagen

Source: Lee Lee.

of the disintegration of the species. As chemicals de-sorb from plastic into the flesh of marine life, the animals experience hormonal changes that cause infertility. Chemicals are invisible, but in the Debris installation plastic serves as a symbolic representation of the chemicals that are harbored within by integrating the material into the corporeal structure of marine creatures. As pieces are reworked then photographed, the images are unavoidably degraded and the creatures obtain a blurred appearance which manifests the undulating quality found in marine environments and bring depth when interspersed with the sharper images. In actual marine environments, growth stems out of decay. Acts of creation and mending balance the destructive method inherent in the process. This creativity is about preservation; a symbolic act of reparation which activates solution building geared towards reducing the amount of plastic pollution at the source.

References

Colborn, T., Dumanoski, D., and Peterson, J. M. (1996). *Our Stolen Future: Are We Threatening Our Fertility, Intelligence, and Survival? A Scientific Detective Story*. New York: Dutton.

Endocrine Disruption Exchange. (2016). Available at: http://endocrinedisruption.org Accessed 6 June 2016.

Freinkel, S. (2011). *Plastic: A Toxic Love Story*. Melbourne: Text Publishing Company.

Humes, E. (2013). *Garbology: Our Dirty Love Affair with Trash*. New York: Avery Books.

Moore, C., and Phillips, C. (2011). *Plastic Ocean: How a Sea Captain's Chance Discovery Launched a Determined Quest to Save the Oceans*. New York: Avery.

Petrini, C. (2015). *Food and Freedom: How the Slow Food Movement Is Creating Change Around the World Through Gastronomy*. New York: Rizzoli International Publications.

Rochman, C. M. (2015). *Marine Anthropogenic Litter; The Complex Mixture, Fate and Toxicity of Chemicals Associated with Plastic Debris in the Marine Environment*. Cham, Switzerland: Springer International Publishing.

Sheller, M. (2014). *Aluminum Dreams: The Making of Light Modernity*. Boston: MIT Press.

Slow Food Movement. (2016). Available at: www.slowfood.com/ Accessed 6 June 2016.

Society of Conservation Biologists. (2016). Available at: http://conbio.org/mini-sites/imcc-2016 Accessed 6 July 2016.

Terry, B. (2012). *Plastic-Free: How I Kicked the Plastic Habit and How You Can Too*. New York: Skyhorse Publishing, Inc.

Terry, Beth. (2016). Plastic-Free: How I Kicked the Plastic Habit and How You Can Too. Available at: www.myplasticfreelife.com Accessed 6 June 2016.

Urry, John. (2013). *Societies Beyond Oil, Oil Dregs and Social Futures*. London and New York: Zed Books.

13 Film Mobilities and Circulation Practices in the Construction of Recent Chilean Cinema

María Paz Peirano

In 2015 the Chilean film *El Club*, directed by Pablo Larraín, won the Berlin Film Festival's Silver Bear, one of the most important world film awards, only the second time a film from Chile had done so.[1] This enormous recognition followed from the growing success of Chilean cinema in recent years in the international film world. Chilean film production, along with its global circulation and exhibition, has considerably expanded, and Chilean films have achieved a wide international recognition directly related to their participation in the film festival circuit. National films have been selected for all the major and most respected festivals, which have also organized forums, talks, and special focuses on Chilean cinema. Films have received a positive critical reception at these events, and have even won various historical awards. Larraín's previous work *No* (2012), for example, was nominated for the first time in Chilean history for the Foreign Language Oscar in 2013, and Sebastián Lelio's *Gloria* (2013) was selected to premiere at the official competition of the Berlinale, the first Chilean film to do so in 20 years. These films' success has helped build the overall prestige of Chilean cinema on the circuits of 'art' and 'alternative' cinema formed by international film festivals, which is the main channel for the circulation, exhibition, and promotion of peripheral world cinema nowadays (Iordanova, Martin-Jones, & Vidal, 2010).

Drawing on an anthropological perspective on film circulation, this chapter discusses how, as with other peripheral film industries, filmmakers' mobilities in the international film festival network have played a major role in this recent upsurge of Chilean cinema. During the last decade, Chilean cinema has emerged thanks to the articulation of both local and global trends, based on the accumulation of symbolic, cultural, and social capital by Chilean filmmakers in transnational settings. Chilean cinema has been positioned on the map of world cinema, due to local transformations which are also connected with international social networks and the economic and cultural exchanges taking place at international film festivals. The shifts in the international circulation and exhibition of Chilean films, and the persistent movements of Chilean filmmakers in these transnational contexts, have entailed new forms of doing and thinking about national cinema, impacting film practices and expanding professional networks beyond national borders.

Filmmakers' Mobilities and Contemporary World Film Production

During the last decade, the global changes in the 'alternative' and 'world' film mediascape have entailed a new focus on the transnational circulation of peoples and films in the film festival network (de Valck, 2007; Wong, 2011). The international film festival circuit has become a relatively stable showcase for peripheral cinemas such as the Chilean one, helping the films reach wider audiences. As suggested by both de Valck (2007) and Falicov (2010), using networked festival spaces as nodes for industrial support seems to be a precondition for the viability of film productions that lack a large domestic market, such as those from Chile. Exhibition at festivals makes films visible at a global level, as their circulation contributes to the value-addition process, and offers possibilities for future distribution. International circulation helps supplement small internal markets, providing an opportunity that is hard to find in Chile, since despite the increase in production levels during the last decade, films do not normally get much distribution at home. Film festivals have also helped support local production, through both direct funding and the possibility of establishing economic exchanges and agreements at the markets, coproduction forums, and other 'industry' spaces hosted by festivals.

A perspective focused on films and filmmakers' mobilities is essential for understanding recent processes of the construction of 'world' cinema (Durovicová & Newman, 2010), that is to say, non-Hollywood peripheral cinema. The mobilities paradigm allows for an interdisciplinary approach, shedding some light on mobile artists' experiences of image-making, film exchange, and circulation. The emphasis on circulation entails drawing attention to the transnational networks shaping these cinemas, as well as to the fluidity of global cinematic production (Cresswell & Dixon, 2002). This paradigm, as well as the multi-sited methodologies it involves, enables us to think of cinema as a social practice that is not fixed to territorial boundaries or 'national art' frameworks, but is embedded in transnational mobile practices and professional networks.

The analysis of contemporary Chilean cinema in this chapter looks at the mobile network of Chilean film professionals involved in creating and exchanging national cinema. This work draws on filmmakers' quotidian mobile experiences on the 'festival circuit,' a network that connects diverse cities around the world, which are turned into nodes for social, economic, and symbolic encounters of professionals, institutions, and films. International festivals encompass not only artistic subjectivities but also urban and national policies and institutional agendas. These nodes can be understood as obligatory 'points of passage' (de Valck, 2007: 38) for the flows in a larger network of film practices, places, and peoples, which depend on the circulation of films and filmmakers. This research looks at this circulation based on multi-sited, 'traveling' ethnographic fieldwork with Chilean filmmakers moving across the international film festival network, particularly the European film festival circuit, between 2012 and 2014. The present case aims to contribute to the still limited empirically based research on filmmakers' mobilities and their role in contemporary cultural production.

According to Stringer (2001), the rise of a mediated urban network of international film festivals has gone hand in hand with the reorganization of the global economy and both the competition and cooperation of cities such as Berlin, Hong Kong, London, and Toronto, which, rather than the national film industries, are now the nodal points in the film circuit. Festival events are part of 'global ecosystems' (Thrift, 2012: 149) constituted by a coalition of the state, local government officials, corporate sponsors, international agents, and nomadic experts, i.e. artists and intellectuals sharing an internationalized cultural capital. Thus, festivals aspire to be hubs for connectivity and exchange in the context of a globalized urban economy promoting international professionalization. As Jamieson (2014) suggests, the rise of both urban festivals and a highly mobile professionalized cultural sector corresponds to the internationalism of the global creative cities paradigm, which has produced boundaries and networks beyond the physical delimitation of the city and the nation. Thus, festival sites have turned into distinctive 'place-specific' events articulating both local and global discourses (Cudny, 2014: 133; Derrett, 2004), which aim to 'mediate cosmopolitanism' (Tzanelli, 2010) and attract film professionals sharing a sense of 'global culture.'

Although both the social and mobile dimension of film festivals are key features for this type of events (Hannam, Mostafanezhad, & Rickly, 2016; Quinn & Wilks, 2013), until recently neither Film Studies nor Anthropology paid much attention to the particular movements and exchanges of filmmakers in the festival circuit. The relationship between contemporary film and spatial mobilities has mostly been discussed considering the nomadic character of certain cinematic representations (see Roberts, 2012), particularly migrant cinema and that from diasporic filmmakers in exile (Rueschmann, 2003; Naficy, 1999, 2001). Only more recently have scholars refocused attention on the international movements and connections embedded in other forms of film production, concerned with the transnational dimension of contemporary non-Hollywood cinema (see Shaw, 2013; Higbee & Lim, 2010; Naficy, 2010), particularly Latin American cinema (Alvaray, 2013; Dennison, 2013).[2] Current studies on film festivals have opened a space to discuss how mobility patterns affect contemporary film production, which is increasingly based on models of international coproduction, funding, and exhibition depending on transnational connections (see Campos, 2015; Falicov, 2010). From a Film Studies perspective, however, these aspects have been mostly addressed regarding the analysis of specific films or corpuses of films, more than the cultural experiences involved in the creation of these films, and the social aspects involved in these forms of production.

An anthropological approach can add to the understanding of this dimension of film production, discussing from a qualitative perspective the role of artists' and artworks' mobilities in their creative processes. The study of the circulation of art and artists follows a long-standing anthropological interest in the exploration of the journeys of objects and peoples across boundaries, noticeably following a discussion of the value-creation process (see Appadurai, 1986). For anthropologists, the role of circulation and exchange in the process of culture making is essential to understanding cultural production, particularly if we see the ways in which objects of art tend to gain value through the same process of exchange, where their 'artistic' quality is also assigned (Myers, 2001, 2004; Clifford, 1988). However, the circulation of films and filmmakers has not yet been studied closely by anthropologists, who have only recently looked at national film production cultures in more detail (Ganti, 2012; Ortner, 2013; Park 2014).

In this chapter we will see how contemporary cinema embodies the material outcomes of the world of mobile social relationships established by filmmakers and other film professionals at film festivals. As suggested by Gell (1999), art is something that emerges in the context of social interaction, incorporating the relationships between peoples and things, and peoples through things. In the case of cinema, I suggest that films are relational objects that are part of the process of reproducing social relations, revealing the impact of social mobilities involved in their creation. The meaning of films is enacted through practices of reception and exchange, articulating global and local trends, and entwined with diverse regimes of value in those settings that shape and promote global film production.

This perspective challenges a prescriptive model of cultures and 'national art' as bounded entities. In turn, film mobilities lead us to think through the fluidity of cultural production, looking at festivals as global 'assemblages' (Deleuze & Parnet, 1987; Müller, 2015) that organize heterogeneous and transnational flows of peoples, films, and narratives. Understanding festivals as 'territorialised assemblages' (Collier & Ong, 2005: 4), where the global intersects filmmakers' subjectivities and creativity, therefore involves acknowledging the permeability of socio-cultural boundaries and the ways in which global movement can affect local filmmaking and national cinemas.

While deconstructing the boundaries of national film production, I argue that contemporary Chilean cinema is increasingly dependent on transnational networks of production and circulation, and on the mobilities of Chilean filmmakers. The national and international articulations discussed below draw attention to the nomadic character of Chilean film practices, and the role of festivals as networked nodes of contemporary cinema. Traveling on the circuit, filmmakers negotiate their creative practices and develop diverse transnational collaborations, a fact that highlights the role of filmmakers' mobilities and festival nodes for the production of national world cinemas.

The Film Festival Network and the Internationalization of Chilean Cinema

Chilean cinema is a 'small national cinema' (Hjort & Petrie, 2007), with an irregular level of production, a small domestic market, and a strong dependence on support from the nation-state. Contrary to big industries, belonging to important world centers of film production with strong internal markets, Chilean cinema is built at the periphery of the global film market. Similarly to other small cinemas, the precarious conditions of production at home and the difficulties of the local market have encouraged the increasing internationalization of local cinema, that is to say, the exchange of films abroad and the acquisition of international funds for the production of national films. Since the mid-2000s, Chilean film professionals have acknowledged the considerable significance of festivals for this internationalization, considering the role they play in enhancing the visibility, distribution, and funding of their films. This renewed interest in international film festivals has coincided with the transformations undergone by these events, which have increasingly opened spaces for international market exchanges under a 'business model' (Peranson, 2009: 25–26) during the last decade. Chileans' attention has been attracted by not only the exhibition possibilities but also the importance of coproduction meetings, training panels, pitching forums, and the like, and they have systematically increased their participation in these events.

International film festivals form a tightly woven network which, taken together and in sequence, form a 'circuit' (Iordanova & Rhyne, 2009). Thus, the network constitutes a cluster of consecutive venues to which film professionals from different parts of the world travel every year, following a set calendar starting in May at Cannes and ending in February at Berlin.[3] This festival network is a dynamic mediascape (Appadurai, 1996) that incorporates overlapping circuits of production, distribution, and consumption, which enable transnational mobilities of images, peoples, and identities. Each festival in the circuit constitutes a global node that works as a marketplace, a competitive venue, and a cultural showcase. At these complex settings, diverse professional and institutional agendas converge, and different agents interact through an overlapping set of international economic and cultural exchanges, grounded in the social dynamics of festival events. Thus, both the festival network and the professionals from small peripheral cinemas like the Chilean one depend to a considerable degree on the several social networks created and strengthened at these sites thanks to these intense interactions.

Chileans have increasingly traveled to international festivals on a regular basis, as a result of these global trends of circulation and the particular transformations of the local field. National transformations are related to diverse changes in Chilean cultural policy, including new legal dispositions, the creation of specialized government institutions, and the increase of state funds for the production, distribution, and promotion of national cinema. The involvement of the Chilean state in the field has meant increasing levels of film production during the last decade linked to a professionalized network of filmmakers working and establishing multiple alliances with other professionals and institutions. In addition, the Chilean government has actively promoted the participation of Chilean cinema in the international circuit since the mid-2000s. A government fund sponsors the trips of film professionals to renowned international film festivals and film markets worldwide, including fiction, documentary, short, and animation festivals. The fund finances travel, accommodation, and other expenses[4] for film festivals all over the world.[5] For the most prestigious and relevant festivals in the international network, film producers have organized the collective Chilean Delegation or 'Chilean Mission,' which comes together at these events. This gathering has turned into an institutionalized practice that links Chileans' disposition to congregate with other co-nationals at these events and the strategy of organizations such as CinemaChile, which put together these missions sponsored by the Chilean government.[6]

The collective use of international festival spaces has helped Chilean film professionals position themselves in the international field of film production, using the label 'Chilean national cinema' to both establish social networks and exchange their films abroad. Filmmakers' frequent circulation through the film festival circuit has also impacted the prestige of Chilean film, and the accumulation of 'festival experience' has increased professionals' social networks, and hence transnational work relationships and cultural and economic exchanges. The value-addition process of films and the accumulation of knowledge and prestige in the field are entwined with the increasing Chilean social capital at transnational settings, drawn from the everyday experience of the festivals. Accumulative experiences help gain personal success in the field since, as suggested by Bourdieu (1993) regarding art elites, accruing social capital in the film world has an impact on the accumulation of cultural capital, which could eventually be transformed into an accumulation of economic capital for filmmakers. I detail below how these processes work for Chilean professionals in the festival network.

The Value of Chilean Films in the Festival Network

In a context in which the global market is dominated economically by Hollywood film production, film festivals are deemed to be showcases of 'alternative' and 'world' (non-English speaking) cinema *par excellence*, for they are the main places where the latest cinema from small and peripheral national industries is exhibited. Festivals constitute commercial hubs for alternative cinemas, and their social structure facilitates the development of particular kinds of peripheral film business.[7] They are sites for the commodification of world cinema, based on the agents' *habitus* and the rules of the film world, which encourage a modernist conception of 'author cinema.' Festivals support this process by reinforcing the meanings and performance of authorship by actively scouting for new talents to exhibit, in addition to funding and training 'emerging' filmmakers—that is to say, international directors providing groundbreaking films (see Ostrowska, 2010).

The artistic value of films is deemed to be the most important value in the festival world, and is the basis of festivals' programming structure. Film selection and awards are based on a criterion of 'quality art cinema,' defined by each festival's curatorial line, which is enacted by authorized gatekeepers such as programmers. Their selective tastes have defined what in the international film world have been called 'festival films' (Wong, 2011: 100): films that conform to certain expectations of artistic quality according to those international standards developed on the circuit. These standards are constructed and renewed by the different agents meeting constantly on the circuit (programmers, critics, producers, filmmakers, etc.). Thus, films' possibilities of being successful and turning into part of a world cinema 'canon' are somewhat intertwined with the social constitution of the field.

Chilean 'festival films' have been able to embody the regimes of value of the international world of art cinema, for Chilean film professionals' practices are very much framed within this notion of author cinema. International festivals have become an ideal place to exhibit and promote Chileans' work, since festivals form a protected niche space where professionals find specialized audiences and secure visibility in the global marketplace. Through their active participation in film festivals, and their selection and awards, Chilean films and filmmakers get global credentials of artistic prestige, and increase their chances of further commercialization. The value-addition process embedded in festival films' circulation means that films, and their filmmakers, increasingly gain symbolic value from their participation in the circuit, which can eventually lead to their accumulating exchange value.[8] Films selected for prestigious festivals are not only 'marked' by their presence but also are more likely to be seen by a larger number of relevant agents: journalists, film critics, and festival programmers, as well as sales agents and buyers. This helps create a 'snowball effect,' increasing their chances of subsequent participation in other festivals and distribution.

As mentioned above, the accumulative value entailed in the circulation of individual Chilean films has also helped to expand the prestige of 'Chilean cinema' as a whole. Chilean films' prestige has transferred to films grouped under the same category, such as 'national' cinemas. The constant circulation also has marked Chilean professionals' trajectories, adding value to their careers by increasing their personal experience and their professional reputation. Their presence at prestigious festivals is both a sign of status and an indication of belonging to the international art-film world—a status that also ensures further circulation in the festival circuit.

Nodes for Transnational Social Networks

In addition to the exhibition of their finished films, directors' and producers' presence opens the door to further social and economic international articulations, facilitating the insertion of other Chilean films in the international market. Festivals are ideal platforms for professionals to prepare the ground for selling unfinished films, 'pitching' and exchanging ideas with key figures in the field, and sometimes getting extra funding for their projects.

'Industry' settings at festivals help secure the attention of the right audience and encounter the most suitable partners.[9] Even when Chilean films do not get immediate revenue, film professionals gain experience, develop new ideas, and meet new people, adding to their own cultural and social capital. 'Knowing each other' at festivals ensures more productive and stable relationships in the field, and it also has an impact on the films showcased at the festivals. For example, filmmakers increase their possibilities of being considered in festivals' programs when they have previously met festival programmers and directors. In turn, a film programmer would normally make the effort to get to know personally the sales agent or the film producer in charge of a film in order to assure its presence in the festival's selection. Film exhibition in each festival edition depends to an extent on previous transnational connections between festival organizers, industry professionals, and filmmakers, whose relationships shape the content of these events. Thus, the films that non-professional audiences get to watch at international festivals are marked by this set of relationships behind the scenes.

The creation of social networks is not achieved just by sharing intense social experiences at one particular festival, but it is the constant re-encounter with other film professionals that enables the creation of reliable relationships. Far from being indelible, international relationships need to be actualized through constant interactions at different festivals. Thus, although each festival is something of a closed world that promotes an intense localized 'communitarian' formation, professionals' mobilities are what assure the permanence of the international social network. Moving across the festival world is a constitutive part of the festival experience, for this constant circulation enables professionals to repeatedly encounter each other and develop social bonds through time.

Film professionals' relationships are reinforced by the expectation of each other's presence, and the reliability of social relations depends on the persistent performances of filmmakers on the circuit. The expectation of the encounter helps foster trust, giving a sense of stability to the social experience, since festival encounters provide continuity throughout the fragmented settings of the festival circuit. Amid the multi-located festival experience, in constant motion, these re-encounters provide a sense of the imaginary permanence of the 'international community.' The time intervals between intense festival gatherings lose their disruptive quality thanks to these recurrent encounters, which provide some consistency to professionals' weak ties.

The above ensures the continuity of the social relationships established there. Despite the fact that different agents have diverse and even competing agendas at the festival, they create cooperative relationships through collaborative practices that come together at these events, thus fostering a sense of belonging to the international film world. The festival experience entails multiple, non-stop social exchanges, which facilitate the creation and maintenance of social alliances that assure economic exchanges in the field. Long-standing social relationships help filmmakers deal with the uncertainty of the film business, constructing bonds that are grounded in a tacit reciprocity. Each festival then turns into a momentary locus of an otherwise de-localized 'international' film world, based on a relatively stable network of people, films, and institutions that is fostered by these agents' constant circulation in the circuit.

Mobile Cosmopolitan Artists

The networking and circulation practices described above have had a significant impact on both filmmakers' subjectivities and the films they make. The 'mobilities paradigm' (Sheller & Urry, 2006) emphasizes how embodied mobile practices are central to how we experience the world (Cresswell & Merriman, 2011) and, in the case of filmmakers, how this experience shapes image-making practices. Chileans' mobilities through the circuit reinforce their internationalist imaginary, that is a certain consciousness of being part of something bigger than only the 'national,' which has enabled them to position themselves as cosmopolitan 'global citizens' who are part of a community of world filmmakers besides the national one.

The cosmopolitan attitude of Chilean filmmakers observed at festivals means that, as part of the 'international community,' they talk on equal terms with other international professionals, while at moments disclosing their national differences, constantly negotiating their position as peripheral filmmakers in the global scene. Chileans' cosmopolitanism is both an attitude and an acquired skill for managing meanings in transnational contexts (Szerszynski & Urry, 2006), and a strategy to mark social inclusion and distinction. On the festival circuit filmmakers behave as 'cosmopolitan subjects' (Hannerz, 1996) that are on the move in the world, strategically engaging and disengaging with local and global discourses. This cosmopolitan attitude is mediated by the international network of film festivals, and promoted by a globalized urban economy, while it also coincides with new practices and imaginaries of mobility in Chilean society, moving away from previous dominant domestic discourses of transnational immobility (Salazar, 2013: 246).

Chilean filmmakers have embraced these new mobility practices as part of the quotidian experience of being a contemporary filmmaker. Most Chilean directors consider themselves both national and international artists, similarly to other 'global' artists (see Schneider, 2006). Without denying their national belonging, they loosen the constraints of being considered exclusively Chilean, and move away from the historical obligation imposed on local filmmakers to produce strictly national cinema. Chileans' aspirations have to do with the project of holding a symbolically universal art-cinema label, beyond the constrained borders of their nation. Their transnational experiences have consequently influenced their creative practices and the aesthetic of their films. Even when their films make reference to certain national traits, and they think themselves as part of a national cinema, films also convey a certain 'cosmopolitan' aesthetic in dialogue with world trends in contemporary cinema. Recent Chilean cinema's aesthetic is common to that of other 'festival films' on the circuit and meets the expectations of quality filmmaking of film festivals, which partly explains their recent success there.

Conclusions

The international circulation of Chilean cinema in the international film festival circuit has had an important impact on national production. I have showed that filmmakers' mobilities and the recurrent transitional exchanges at these sites have enabled the creation of networks that facilitate global film circulation, and that the festival experience has also shaped the production and exhibition practices of Chilean filmmakers. The professionalization of recent Chilean film production has also been grounded in the intensive participation of Chileans in the film festival circuit. Festivals have marked professionals' recent trajectories, changing some of their perceptions of both the film world and the possibilities for their own filmmaking, including ways to make, fund, distribute, and promote their films. The circulation of film professionals in the marketplace and their feeling of belonging to the international, cosmopolitan film community have also marked artists' subjectivities and films.

Mobile social practices and configurations have had an important impact on the social networks that make possible both the existence and the regimes of value of recent national cinemas. As shown in the Chilean case, mobile festival experience has structured filmmakers' practices, shaping the ways in which they position themselves in the international field of alternative world cinema. Festivals, understood mainly as social hubs entwined with narratives and regimes of value of independent art cinema, provide the space for the transnational encounter between peripheral filmmakers and the key figures in the field of world cinema. These networked urban nodes allow for the encounter of multiple agents, which shape the field and connect peripheral film professionals with the rest of the international film world. Both Chilean filmmakers' and films' mobilities through the festival circuit have helped them create transnational social networks, exchange their films, and thus accumulate the necessary symbolic, social, and cultural capital in order to keep participating in the international field. Likewise, their 'national' film production transcends national boundaries, demonstrating the transnational connections embedded in contemporary world cinema production.

Notes

1. The only other film to have previously won this award was *La Frontera* (The Frontier, 1991) by Ricardo Larraín, which was also the only Chilean film selected for the official competition of the Berlinale until 2013.
2. This relationship has also been explored more recently in relation to Film Tourism, see for example Connell (2012), Tzanelli (2007). For more recent and extended analyses of transnational film, see also the journal *Transnational Cinemas*, which has focused on this topic since 2010.

3. There is no agreement between Cannes and Berlin regarding which one officially starts the calendar, since both claim their privileged position to open the circuit. Given that Cannes' *Marché du Film* is still predominant in the international market, I consider it more adequate to stick to this starting point here.
4. The fund includes the following: a plane ticket, accommodation for one person, Digital Cinema Package (DCP) master for exhibition, subtitles for the copy participating at the festival, sending material to the festival, and also materials and resources to buy exhibition spaces at the festival. Source: CAIA (2014).
5. The fund sponsors Chilean participation at Cannes (France), Berlin (Germany), San Sebastián (Spain), Toronto (Canada), and Sundance (USA). It also sponsors attendance at other very important film festivals such as IDFA (Amsterdam), Guadalajara (Mexico), La Habana (Cuba), Clermont-Ferrand (France), BAFICI (Argentina), Toulouse (France), Visions du Réel (Nyon), Hot Docs (Canada), DokLeipzig (Germany), Annecy (France), and Karlovy Vary (Czech Republic).
6. CinemaChile is a private agency that promotes Chilean cinema abroad, and is linked to the *Asociación de Productores de Cine y Televisión* (APCT, Association of Cinema and Television Producers).
7. The most important festivals are increasingly turning into business-oriented events in order to maintain their existence. This has been the cause of much criticism from some film critics and international cinephiles, who see in this turn the displacement of the focus on 'art cinema' and a threat to festivals' traditional artistic and political aims (Peranson, 2009).
8. For a detailed analysis of this process regarding other peripheral cinemas, see also Kocer (2013), Iordanova and Rhyne (2009), and de Valck (2007).
9. See Vallejo (2014) for an extended analysis on the role of industry sections for other small cinemas, particularly for documentary filmmakers.

References

Alvaray, L. (2013). Hybridity and Genre in Transnational Latin American Cinemas. *Transnational Cinemas*, 4(1), 67–87.

Appadurai, A. (1986). *The Social Life of Things: Commodities in Cultural Perspective*. Cambridge: Cambridge University Press.

Appadurai, A. (1996). *Modernity at Large: Cultural Dimensions of Globalization*. Minneapolis: University of Minnesota Press

Bourdieu, P. (1993). *The Field of Cultural Production: Essays on Art and Literature*. Cambridge: Polity.

CAIA, Consejo del Arte y la Industria Audiovisual. (2014). *Fondos de Cultura: Bases para concurso Fondo de Fomento Audiovisual, Programa de Apoyo para la participación en Festivales y Premios internacionales*. Santiago: Consejo Nacional de la Cultura y de las Artes. Available at: http://chileaudiovisual.cultura.gob.cl/ Accessed 20 May 2014.

Campos, M. (2015). Film (Co)Production in Latin America and European Festivals. *Zeitschrift Für Kulturmanagement*, 1(1), 95–108.

Clifford, J. (1988). *The Predicament of Culture: Twentieth-Century Ethnography, Literature, and Art*. Cambridge, MA: Harvard University Press.

Collier, S. J., and Ong, A. (2005). Global Assemblages, Anthropological Problems. In A. Ong, and S. J. Collier (Eds.), *Global Assemblages: Technology, Politics, and Ethics as Anthropological Problems* (pp. 3–21). Oxford: Wiley-Blackwell.

Connell, J. (2012). Film Tourism—Evolution, Progress and Prospects. *Tourism Management*, 33(5), 1007–1029.

Cresswell, T., and Dixon, D. (Eds.) (2002). *Engaging Film: Geographies of Mobility and Identity*. Washington, DC: Rowman & Littlefield Publishers.

Cresswell, T., and Merriman, P. (2011). *Geographies of Mobilities: Practices, Spaces, Subjects*. Farnham: Ashgate.

Cudny, W. (2014). Festivals as a Subject for Geographical Research. *Geografisk Tidsskrift-Danish Journal of Geography*, 114(2), 132–142.

Deleuze, G., and Parnet, C. (1987). *Dialogues*. New York: Columbia University Press.

Dennison, S. (Ed.) (2013). *Contemporary Hispanic Cinema: Interrogating the Transnational in Spanish and Latin American Film*. Woodbridge: Tamesis.

Derrett, R. (2004). Festivals, Events and the Destination. In Y. Ian, R. Martin, Jane Ali-Knight, D. Siobhan, and Una McMahon-Beattie (Eds.), *Festival and Events Management: An International Arts and Culture Perspective* (pp. 32–51). Amsterdam: Elsevier Butterworth-Heinemann.

De Valck, M. (2007). *Film Festivals: From European Geopolitics to Global Cinephilia*. Amsterdam: Amsterdam University Press.

Durovicová, N., and Newman, K. (2010). *World Cinemas, Transnational Perspectives*. New York: Routledge.

Falicov, T. (2010). Migrating from South to North: The Role of Film Festivals in Funding and Shaping Global South Film and Video. In G. Elmer, C. Davis, J. Marchessault and J. McCullough (Eds.), *Locating Migrating Media* (pp. 3–21). Lanham: Lexington Books.

Ganti, T. (2012). *Producing Bollywood: Inside the Contemporary Hindi Film Industry*. Durham: Duke University Press.

Gell, A. (1999). *The Art of Anthropology: Essays and Diagrams*. Oxford: Berg Publishers.

Hannam, K., Mostafanezhad, M., and Rickly, J. (Eds.) (2016). *Event Mobilities: Politics, Place and Performance*. London: Routledge.

Hannerz, U. (1996). *Transnational Connections: Culture, People, Places*. London: Routledge.

Higbee, W., and Lim, S. H. (2010). Concepts of Transnational Cinema: Towards a Critical Transnationalism in Film Studies. *Transnational Cinemas*, 1(1), 7–21.

Hjort, M., and Petrie, D. J. (2007). *The Cinema of Small Nations*. Edinburgh: Edinburgh University Press.

Iordanova, D., Martin-Jones, D., and Vidal, B. (2010). *Cinema at the Periphery*. Detroit: Wayne State University Press.

Iordanova, D., and Rhyne, R. (2009). *The Festival Circuit*. St. Andrews: St. Andrews Film Studies.

Jamieson, K. (2014). Tracing Festival Imaginaries: Between Affective Urban Idioms and Administrative Assemblages. *International Journal of Cultural Studies*, 17(3), 293–303.

Kocer, S. (2013). Making Transnational Publics: Circuits of Censorship and Technologies of Publicity in Kurdish Media Circulation. *American Ethnologist*, 40(4), 721–733.

Müller M. (2015). Assemblages and Actor-Networks: Rethinking Socio-material Power, Politics and Space. *Geography Compass*, 9, 27–41.

Myers, F. (2001). *The Empire of Things: Regimes of Value and Material Culture*. Oxford: James Currey.

Myers, F. (2004). Ontologies of the Image and Economies of Exchange. *American Ethnologist*, 31(1), 5–20.

Naficy, H. (1999). *Home, Exile, Homeland: Film, Media, and the Politics of Place.* New York and London: Routledge.

Naficy, H. (2001). *An Accented Cinema: Exilic and Diasporic Filmmaking.* Princeton and Oxford: Princeton University Press.

Naficy, H. (2010). Multiplicity and Multiplexing in Today's Cinemas: Diasporic Cinema, Art Cinema, and Mainstream Cinema. *Journal of Media Practice*, 11(1), 11–20.

Ortner, S. (2013). *Not Hollywood: Independent Film at the Twilight of the American Dream.* Durham: Duke University Press.

Ostrowska, D. (2010). International Film Festivals as Producers of World Cinema. *Cinema & Cie*, 10(14–15), 145–150.

Park, Young-a. (2014) *Unexpected Alliances. Independent Filmmakers, the State and the Film Industry in Postauthoritarian South Korea.* Palo Alto: Stanford University Press.

Peranson, M. (2009). First You Get the Power, Then You Get the Money: Two Models of Film Festivals. In R. Porton (Ed.), *Dekalog 3: On Film Festivals* (pp. 23–37). London: Wallflower.

Quinn, B., & Wilks, L. (2013). Festival Connections: People, Place and Social Capital. In G. Richards, M. De Brito and L. Wilks (Eds.), *Exploring the Social Impacts of Events* (pp. 15–30). London: Routledge.

Roberts, L. (2012). *Film, Mobility and Urban Space: A Cinematic Geography of Liverpool.* Oxford: Oxford University Press.

Rueschmann, E. (Ed.) (2003). *Moving Pictures, Migrating Identities.* Jackson: University of Mississippi Press.

Salazar, N. B. (2013). Imagining Mobility at the "End of the World". *History and Anthropology*, 24(2), 233–252.

Schneider, A. (2006). *Appropriation as Practice: Art and Identity in Argentina.* New York: Palgrave Macmillan.

Shaw, D. (2013). Deconstructing and Reconstructing 'Transnational Cinema'. In S. Dennison (Ed.), *Contemporary Hispanic Cinema: Interrogating Transnationalism in Spanish and Latin American Film* (pp. 47–66). Woodbridge: Tamesis.

Sheller, M., and Urry, J. (2006). The New Mobilities Paradigm. *Environment and Planning*, 38(2), 207–226.

Stringer, J. (2001). Global Cities and International Film Festival Economy. In S. Mark and F. Tony (Eds.), *Cinema and the City: Film and Urban Societies in a Global Context* (pp. 134–144). Oxford: Blackwell.

Szerszynski, B., and Urry, J. (2006). Visuality, Mobility and the Cosmopolitan: Inhabiting the World from Afar. *The British Journal of Sociology*, 57(1), 113–131.

Thrift, N. (2012). The Insubstantial Pageant: Producing an Untoward Land. *Cultural Geographies*, 19(2), 141–168.

Tzanelli, R. (2007). *The Cinematic Tourist: Explorations in Globalization, Culture and Resistance.* London: Routledge.

Tzanelli, R. (2010). Mediating Cosmopolitanism: Crafting an Allegorical Imperative through Beijing 2008. *International Review of Sociology*, 20(2), 215–241.

Vallejo, A. (2014). Industry Sections. Documentary Film Festivals Between Production and Distribution. *Illuminace: Journal of Film Theory, History, and Aesthetics*, 26(1), 65–82.

Wong, C. (2011). *Film Festivals: Culture, People, and Power on the Global Screen.* New Brunswick: Rutgers University Press.

14 The Roberto Cimetta Fund as a Response to Artistic and Cultural Mobility Imbalance

Angie Cotte

Artistic and cultural mobility is a new-old paradigm that has been neglected by scholars in the past but which is today proving to be a crucial concept in unraveling some of the knots that prohibit global cultural relations. Artistic and cultural mobility refers to individual transnational travel for cultural and artistic reasons. In just a few decades, obstacles to this mobility—visa restrictions, lack of funding, border controls—have increased in many parts of the world, notably across the Euro-Arab divide, but also between Arab countries or between Asian countries and countless other areas; often due to country-to-country disputes, political embargoes, civil repression, conflicts, and so on. These impediments to mobility have weakened artistic and cultural exchanges. From our often Eurocentric viewpoint we neglect this important fact, because for Europeans, mobility has become easier in the context of the European Union. As Najwan Darwish, a Palestinian writer and poet recently explained at the 'Istikshaf Symposium' in Alexandria in June 2012, "the Arab world is living [in] the dark ages of mobility."

The Roberto Cimetta Fund (RCF) was set up in 1999 to respond to these issues by providing travel grants to and from Europe and the Arab countries. The Barcelona Process had only just been set up in 1995 to establish Euro-Mediterranean policies and programs for the heritage and audiovisual sectors but without any available funding for the contemporary arts in general. RCF's secretariat was set up in Brussels by European cultural operators and funded by the European Cultural Foundation and a few other donors as a response to this mishap. Over the span of ten years the Fund was able to provide approximately 100 travel grants a year for actors, theater directors, choreographers, and dancers to travel to and from both zones in order to perform, take part in residencies, prepare projects, and attend training courses.

During that period (from 1999 to 2009) the obstacles to mobility in the Arab region did not improve. The Arab Charter of Human Rights (articles 20, 21, and 22), which foresaw the possibility to move from state to state and re-enter one's own territory, became restricted in 2004 to only one article (article 26.1) that gave the right to move within one's own territory without any mention of travel from state to state (see Figure 14.2). In some

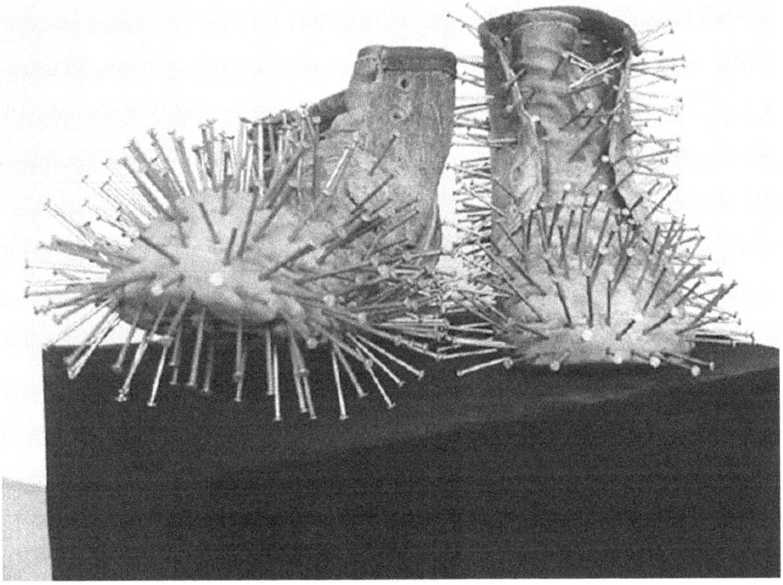

Figure 14.1 Illustration by Mohammed Al Hawajri—Cactus Borders project from 2005

Source: Permission from the artist.

Arab countries a person that is known to perform (as an actor or dancer) or who produces video or cinematographic art will be forbidden to travel; similarly a woman traveling alone will be asked to be escorted by a father, brother, or husband. Requesting a visa in these countries sometimes entails paying a substantial fee, informing the authorities of the income one has in one's bank account, and/or exposing oneself to humiliating procedures and preconceptions from civil servants about motives to migrate rather than to travel for cultural or artistic reasons. Mohammed Al Hawajri, Palestinian artist and RCF grantee (Figure 14.1, above), had to wait three months before returning to Palestine because of laws that had changed in his home country during the time he was abroad. Similarly, artists wishing to prepare international artistic projects in the Arab region may come across many difficulties to set up their project and find sufficient funds. Nesrine Chaabouni wanted to set up the Tunisian section of the International Dance Council powered by UNESCO but only found a travel grant from the RCF to enable her to attend the World Congress in Athens in 2014 so that her project could become reality. There are countless stories of difficulties to travel and

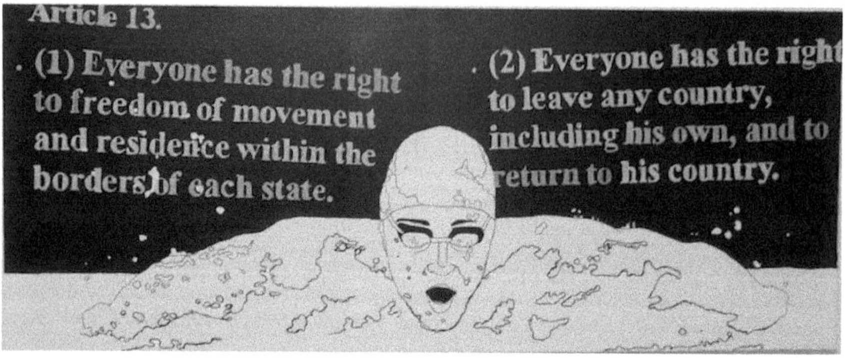

Figure 14.2 Work by Oussama Tabti, Article 13 of the Declaration of Human Rights, 2012

Source: Courtesey of Oussama Tabti.

circulate which have been collected by the Safar Fund in Jordan (www.safarfund.org), who are the lead partner of the Istikshaf platform, of which the RCF has been a co-partner for the last four years (see Istikshaf Facebook Campaign).

This rather bleak situation has not deterred the RCF from its task. From 2010 onwards, under the dynamic impulse of Ferdinand Richard,[1] newly elected Chair of the Fund, new funding partnerships were sought (12 new institutional partnerships from seven different countries including Iraq to date), the Board of Directors was enlarged to include both Arab (and Turkish) and European professionals, the Expert Committee which evaluates the travel grant applications was increased to 35 members from all artistic fields and all countries of the Euro-Arab geographical zone, applications were made available to cultural operators as well as artists, and Arab artists or cultural operators wishing to travel outside of the Euro-Arab zone to other parts of the world were also allowed to obtain funding. RCF was particularly attentive to the link between the Arab and Balkan regions because both regions suffered from isolation due to mobility obstacles; we thus organized two events in Belgrade (2013) and Zagreb (2014) to encourage networking between artists and cultural operators of these regions. Many Arab operators had never been to Zagreb or Belgrade before.

Current imbalances in art production flows persist because of funding problems. Indeed, Arab artists are often invited to take part in European projects or projects funded by the EU. When they return home there is no 'local' money to set up their projects in a sustainable way and they cannot make use of the benefits obtained through their travel abroad. These artists will want to return to Europe where opportunities are available for them, which contributes to the brain-drain effect, weakening the cultural fabric in the South. Cultural policy reform should address the issue of how to make their local territories attractive to their own youth, not only to tourists. The heavy investments local authorities provide in the education of creative young adults at a local level would thus have a concrete 'pay-back' effect if considered from this standpoint.

RCF undertook important evaluations and built up collective-intelligence working with cultural and arts operators in these regions (350 RCF grantees, over 2000 local and regional arts and culture organizations in the Euro-Arab geographical zone). Following two recent mobility impact assessments carried out by RCF's office in 2012 and 2013 we were able to determine more specifically some key characteristics of artistic mobility. Firstly, the artistic travelers consider that the most important role of cultural and artistic mobility is to develop understanding of the world and globalization and promote understanding between different cultures.

Their personal motivation to travel is related to the concrete productive artistic outcome of their trip and a better integration into their professional circuits. They realize the impact of their trip very quickly and are able to qualify it as an experience that they want to pursue through further mobility and by keeping contact with the people they have met. In general 60% of travelers want to renew their mobile experience and over 90% stay in touch with their hosts or the people they have met during their stay abroad. Not only has the experience been generally positive but it has helped them contextualize the issues in a given region. This experience invariably widens their perception, giving them new insight on their home locality when they come back and enabling them to see other possibilities that they hadn't seen before.

Perception of cultural differences between home and host help them to break down certain mental barriers related to aesthetics, ideologies, or politics, developing new ideas and approaches that open up a certain freedom of thought and movement that is beneficial to their work. Often, by meeting other artists from a variety of nationalities, mobile artists can envisage their projects developing with many different people in different ways, so that the coproduction or exchange can be envisaged multilaterally.

It can happen that a mobile experience doesn't work out so well. Reasons for this stem from lack of preparation with the host organization or lack of an intermediary (i.e. person who puts you in touch with people interested in your project). Administrative complications can arise while the artist is abroad or during the return trip which can entirely damage the project and its outcomes. Due to the stricter border controls in the last several years, artists are finding digital ways to make their work travel and be exhibited.

Being mobile and productive cannot be possible without intermediaries or relays (i.e. organizations that already exist and help artist to find other contacts in other countries). RCF acts as a funding intermediary between institutions and individual artists. RCF is also an information intermediary linking misguided artists to other sources of information, as well as an event-organizer, enabling artists to meet, strengthen their capacities, and build project partnerships and contacts. Other intermediaries are, for example, networks of operators that organize events so that the sector can interact at European or international level (such as the Istikshaf platform or Arterial network, for example). They can also be platforms of exchange in which operators come together at a given moment on a given topic to build up a framework for action, for project-building, or to advocate for their needs (see Transeuropehalles network). Translation is also a necessary tool for mobility to further the capacities of artists to communicate and translate their own works (see also the work done by Transeuropéennes). Information needs to circulate to all actors so that opportunities, financial resources, events, regulations, studies, and so on are available in all countries to all actors wherever they may be (such as *Universes in Universe*, *Babelmed*, *Culturelink*, *Interartive Magazine*). Representation of the cultural and arts sectors is vital to coordinate the needs, advocate, or relay policy issues at a wider level (such as the platform Culture Action Europe). Strengthening of capacities in the sector through training at international level is also required.

On the basis of our recent studies as depicted above, the wider social, cultural, political, and economic implications of cultural or artistic mobility can be more clearly defined along the following lines:

- The concrete art production resulting from the mobility of artists clearly impacts on local development in the home and host countries and creates a chain of positive effects, linking local art production to a more international pool of entrepreneurs and activists.
- The mobility that RCF supports enables meetings face to face which are the condition of trustworthy, long-term partnerships that can develop in a multilateral way.
- Mobility is a 'two-way' phenomenon that should be supported as such, since by exporting well a country can also receive better and therefore export more—as a chain of collective benefits. Indirectly these mutual benefits encourage reciprocity and mutual respect.
- Building cultural policies around mobility helps to define geographical directions of artistic production, linking artistic hubs which are not necessarily in major urban areas, but often link the inter-regional or inter-local levels. Generating this type of production circuit contributes to renewed and sustainable models of alternative cooperation, diversifying cultural and arts productions, maintaining local markets, local productions, and therefore cultural diversity, thus going against the distribution and production monopoles which are currently damaging the economic stability of small companies, producers, and entrepreneurs from the arts world.

Mobility and the circulation of works is very much about meeting people face to face and building up sufficient confidence and affinities to enable collaboration or even co-created networks of coproduction—interconnecting local actors at an international level. Aesthetics and artistic work develop as a process of experimentation in which each partner's own interculturality is the central working tool. The results of these coproductions are often interdisciplinary, or flexible in form, varying in relation to the audience to be reached (i.e. 'national' or 'local' audiences). An Italian theater director can produce a performance following a workshop with local artists in Palestine and show the performance in two different forms to the local audience and to the audience back in Italy.

Mobility can also entail an absence of hierarchy in the structure of the joint-project. This absence of hierarchy, due to the experimental form of the joint-adventure, often develops forms of small collaborations, involving a relatively small number of collaborators. These micro-groups also develop because they are working on micro-funding models, which must be small and flexible to survive in the absence of proper international funding mechanisms. A new framework of references should then emerge. This framework should first of all consider that mobility is not the aim itself. Equal access and facility to mobility as a way of developing arts and culture in the four corners of the globe is the main goal. The aim is also to increase the autonomy and emergence of artists and their producers.

Mobility is a global policy issue. As such it is a cultural policy issue for Europe and the Arab League, for member states and for local governments. In this context we aim for mobility to be understood as a round-trip concept, directly linked to local development at the starting point. One-way trips from one corner of the world to another are a human rights issue more than a cultural policy/development issue, even if this issue is a central pillar of democracy building. We must acknowledge the geographical imbalances at work between the rural and the urban, between the North and South, the directional imbalances are directly related to colonial heritage. There is an urgent need for a more reciprocal and respectful exchange between Europe, its neighbors, and the world at large particularly the so-called emerging countries. This exchange must also take stock of the need to rebalance the artistic needs such as infrastructures, training, and human resources. Artistic imbalances are not linked to talent and artistic capacities being higher in some parts of the world than in others: Artistic potential exists in any one of us, everywhere. Potential to overcome artistic imbalances depends on the setting up of policies related to given sectors that respond to the operators and artists working at the local level. Building up democratic platforms to address the needs and ideas and transform them into concrete policies in various sectors of artistic activity remains a key issue in Arab countries as well as on the European continent. Are we ready to think in terms of an artistic community without frontiers, which exists through encounters, works through experimentation, and is productive through interaction of a multiplicity of actors—not only from different nationalities but also working at different levels (artists, managers, critics, programmers)?

Some interesting reports have been developed in past years, notably the study entitled 'Mobility Matters' which gave an extensive picture of mobility needs in each European country. Impediments to mobility is also a significant report written by Richard Polacek which clearly outlined the obstacles to mobility so that institutions, the European Commission in particular, could take stock of the situation and act accordingly. IETM, the international network for the performing arts, published a number of studies on this issue showing how crucial the question of mobility was to the production and distribution of contemporary works. RCF's members (some 42 independent, active, and engaged cultural and arts professionals living and working in the Euro-Arab geographical zone) are developing interesting theories on various topics relating to mobility, including how mobility dynamics should be understood in postmodern theory.[2] Ferdinand Richard in particular has written a number of important and challenging articles on these topics, which are available to download on the Roberto Cimetta Fund's website (www.cimettafund.org).

What are we doing today? Thanks to RCF's independent and flexible nature it has been able to accomplish much in a small amount of time. Over the last few years we have been concentrating on developing ties with other independent mobility funds based elsewhere; such as Arts Network Asia, the Russian Theatre Union, Safar Fund, Art Moves Africa, CEC Arts Link, India for Transformation, and others. The aim of this collaboration is to communicate our mobility funding calls at the same time in order to give greater visibility to this issue and at the same time give artists the opportunity to identify the different funding options. All of these organizations are independent non-profit agents that establish institutional funding partnerships and then re-direct the funds to individuals. We believe that such funding mechanisms are necessary in order to respond to artistic and cultural needs in a context where many institutions have not yet developed the appropriate tools to respond to these needs.

What is the context today that is affecting artistic and cultural mobility? As Ferdinand Richard recently pointed out in an Artistic Mobility Alliance meeting in Athens (Figure 14.3),[3] currently the leading global cultural industries are monopolizing the market due to a complete revolution in the nature and function of the *cultural object* and in the nature of distribution channels. These industries attack the diversity of local markets, which are seen as fragmenting the global market; monopolization is for these industries the best way of making business. The strategies of these new bulldozers are counting on rapid capital circulation and short turnover of products. They neglect small, highly specialized markets; these do not appeal to their eager appetites in market shares and instant benefits. Cultural assets and tangible benefits can be found in such specialized markets. Cultural engineering, capacity building, core art skills, social dialogue, bottom-up mushrooming, are developing in these markets and are central elements of cultural or artistic production, not yet controlled by these new massive operators. Cultural

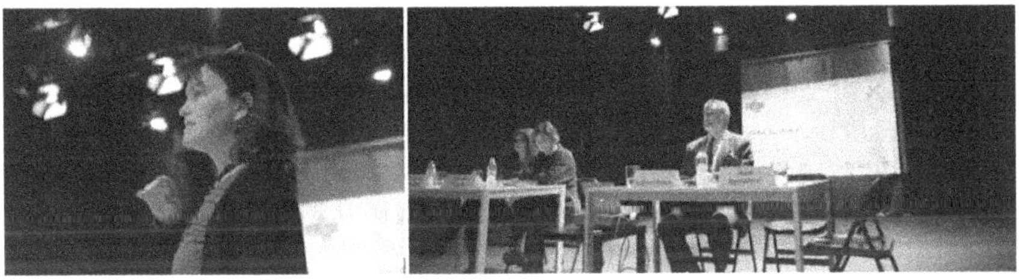

Figure 14.3 Artistic Mobility Alliance meeting in Athens, March 2015, supported by the Asia-Europe Foundation

Source: Photos by Marina Tselepi.

micro-businesses incubators, micro-funding schemes, shared co-working spaces, fablabs, local institutional networks to support cultural start-ups, direct online markets, short distribution circuits, peer-to-peer exchanges of all kinds, are processes that lead to a dignified, independent, autonomous, diversified free market, to freedom of speech and peacekeeping, which are key to any business even those of leading global industries.[4]

At another level, local authorities in Europe have been investing in cultural practitioners and agents through educational and cultural policies for a number of decades now, developing skills, arts production, arts education, and creativity. However, the economic potential and attraction of major cultural capitals is draining cultural professionals away from the local level to major hubs around the world. This 'exporting' of artistic talent and cultural management skills evidently represents a loss in investment for local authorities. Artistic and cultural mobility should not be considered as a one-way ticket outside of the local community but as a return ticket, directly linked to local development at the starting point. Seen from this angle, artistic and cultural mobility is also a major asset in peace building between *cultural subjects* since it brings understanding of world regions and cultures back to the local level.

New models are already developing in this regard particularly in the Arab, Balkan, and Indian regions through innovative organizations such as Kulturanova Foundation in Croatia, Zoomaal in Lebanon, or India for Transformation. Platforms of exchange such as Istikshaf and Cimetta Forum are interconnecting mobility actors and their concerns in the field of small and medium sized enterprises as well as civil society activism. Although cultural entrepreneurs are constantly developing innovative strategies for market-based survival, the very nature of Art means that art practices will always be outside the frame, in a niche, or marginal, not conforming to current market paradigms. For the Artistic Mobility Alliance that RCF is setting up, it is essential to support the artistic quality of a travel project and to evaluate this quality via expert evaluation committees such as the one set up within the RCF.

Mobility funding is part of structural funding contributing to a holistic approach to local art practice development and sustainability facilitated by a diversity of actors from local chambers of commerce, local authorities to clusters or collectives, to new platforms of monetary exchange and participation such as crowd funding. In this context the accountability tied to the funding process can be ensured via virtual platforms that build up local communities of support/access audiences and provide visibility as well as via cultural indicators that measure local cultural development as the concrete outcome of arts mobility.

Our Artistic Mobility Alliance is also concerned with mobility of artists in conflict zones. The Arts Mobility Alliance members must develop an understanding of the ethics they share particularly in unstable environments in which the volatile circumstances at work in these conflict zones can lead to traumas and unhealthy situations on the return. Vasl art collective in Pakistan has learned to become resilient in a context of target-killing, repression, and danger, believing that the ties it develops with Bangladesh and Sri Lanka are key in conflict-resolution processes. This local cross-border mobility and cooperation is as important as international mobility. Establishing ties with local neighbors through local diplomacy is often key in unlocking tricky, long-standing divides that national diplomacies fail to solve.

Funders can perceive the interest of funding such mobility since it provides artistic and cultural insight for them in a difficult territory. This insight allows for a cultural diagnosis in the zone, which can lead to further concerted mobility and development funding based on the analysis provided from the terrain. Scholars still need to investigate areas where knowledge is lacking, such as country-based cultural diagnosis and mobility/funding needs in Egypt, Algeria, and other Mediterranean countries. This information is necessary in order to defend the culture of these countries and devise cultural policies accordingly. The Association Racines in Morocco, the Cultural Agenda in Lebanon, and Ettijahat in Syria are some of the remarkable Arab cultural networks and organizations that are taking up this challenge. Cultural mapping in this zone should be undertaken urgently in order to focus on identifying art venues in localities within the Euro-Arab geographical zone which engage in international artistic exchange. The relevant support frameworks provided by local authorities should also be mapped. Studies related to diaspora need to be undertaken particularly on how diaspora envision their investments for their culture. Finally the notion of re-establishing commonality,[5] which is at the basis of the Cimetta Forum, set up by RCF since 2014, is a complementary study to mobility issues.

The universal declaration of human rights starts by stating that we are all equal in dignity and rights. So the basis of any ethic linked to artistic mobility should be that there is no dominant culture. We should advocate for fair culture, which is related to fair trade, in which the mobility conditions of artists and cultural operators from emerging countries should be respected and where a sustainable approach to partnership building is promoted.

We should advocate for liberty of artistic expression in all parts of the globe and for the independence of the producer, the maker, and the creator of art or artworks wherever he or she may be. Finally and most importantly, the role of culture in conflict resolution and reconciliation should be recognized and supported as such.

Notes

1. See Ferdinand Richard's biography on the RCF website.
2. Vahid Evazzadeh, RCF Board member, paper entitled 'From Negotiated Order to Freedom Proper.'
3. Mobility Funding Alliance report, Athens 2015, https://artisticmobilityalliance.wordpress.com/2015/05/06/report-first-meeting-of-mobility-alliance-partners/.
4. See Ferdinand Richard's writings on global cultural paradigms, www.cimettafund.org/article/index/rubrique/1/lang/en/fm/1/id/1.
5. See Giusy Checola's research at Archiviazioni, Bologna, Italy.

15 Mobile Performing Arts

Facts, Figures, and What They Say About Reality

Bart Magnus

"Not everything that can be counted counts, and not everything that counts can be counted." Albert Einstein once framed it very accurately. It speaks well for him that, as a physicist, he also understood the limits and shortcomings of numbers, formulae, and data to gather knowledge about the world and mankind. It is a noteworthy quote for every researcher working with data to bear in mind when telling great stories based on big data.

On the other hand, and despite rapid developments in recent years, the potential of data research is often not yet fully exploited. In their 2013 paper 'Counting What Counts. What Big Data Can Do for The Cultural Sector,' Anthony Lilley and Paul Moore hit the nail on the head when stating that "too often . . . the data is inadequately collected, left in raw, unanalysed form and/or not brought into play in strategic decision-making" (Lilley & Moore, 2013: 11).

The cultural sector is no exception to this analysis. It is the combination of the two above-mentioned thoughts that should resonate in our heads when talking about data on mobility in the performing arts.

Performing Arts Data Collection in Flanders

Flanders Arts Institute (www.flandersartsinstitute.be) has gained some experience in the topic through more than 20 years of data gathering and research. Since 2015 Flanders Arts Institute serves the arts sector in Flanders, Belgium. It is a merger of VTi (Institute for the Performing Arts in Flanders), BAM (Flemish Institute for Visual, Audiovisual and Media Art), and Flanders Music Centre. Core functions include a focus on research, international activities, supporting the practice of art, and policy support. On a daily basis, knowledge and expertise about and for the arts in Flanders is collected and distributed in an international context. The institute is the ideal contact point for foreign art professionals in search of information on the arts in Flanders. Information on relevant research, directions, and trends in the Flemish arts sector are provided in a tailored way. Visitors' programs are organized to let the international community discover up-and-coming names and must-sees.

Gathering 'facts' about the contemporary performing arts practice is indispensable in the research and information roles that we play. Flanders Arts Institute, and its predecessor VTi, has been monitoring the performing arts field by describing productions and casts with their premieres and tours abroad in a database. By having gathered this information for over a period of more than 20 years, we stand on solid ground to track evolutions.

The sector's interest in culture mobility and working and touring internationally has been growing since the 1980s. This made the international aspect of performing arts productions become one of the key topics in our research. Since the beginning of the data collection, foreign companies or arts organizations coproducing the work of Flemish artists have been added to the database. In 2000, we started collecting information about Flemish productions touring abroad. This perspective is the most obvious and straightforward way of gathering data about international mobility in the performing arts. Contrary to any non-live art form, a work in itself cannot be mobile. In some way it doesn't even exist unless it is performed. Performing arts need artists who perform and who should thus be mobile along with the productions. It would be interesting to investigate how being mobile affects artists and companies. Artistic mobility may have consequences on the way they work and produce and sell their work.

Our data make it possible to answer questions regarding the international nature of performances on both the production side and the distribution side. In 2002, a first report on import and export of productions was published. A first field analysis was published in 2007, followed by a second and third edition in 2011 and 2014. In these field analyses, the distribution side (international touring) and production side (international coproductions) were thoroughly analyzed at the hand of the collected data. What share of Flemish productions leads an international life or is international by nature? Where do these productions tour? Can we see evolutions over time and can they be traced back to any shifts in how artists and companies work or to any actions taken on a local or international cultural policy level? And if so, should we speak about correlation or is there any direct causality? In other words: What do the figures say about reality and how does that complex reality affect the numbers?

But our questions could go even further. Could we for example say that artistic mobility enriches the artistic experience? And if so, does that count for the artist only, or also for the audience? Does artistic mobility enhance in some way the value of the arts for society at large?

Data-Driven Narratives Originating From Flemish Performing Arts Data

Performing arts mobility doesn't happen in a vacuum. It is continuously affected by direct factors (e.g. the sector's self-development and self-organization, culture policy measures on different levels) and indirect factors (e.g. the macro-economical situation). When we take a look at the number of 'Flemish' productions, we see an increase over the years from 481 productions in 1993 to 842 twenty years later. A potentially significant point here turns out to be the financial crisis of 2008, with quite a dramatic drop of productions from which we only recently seem to have recovered. To try to declare this dip, we should dig to a second layer of data. Let's have a look at this production reality from an artistic mobility point of view.

The top line of Figure 15.1 indicates the percentage of productions that tour abroad. The line below indicates the share of international coproductions (with at least one Flemish partner). Again, the season 2008–2009 is a critical point in the interpretation of these data. After this season, we see a rather harsh and unattended fallback in internationally touring productions. At the same time, the share of international coproductions rises faster than ever before (a climb of 7% in two years' time). Combining quantitative and qualitative research could turn out to be elucidating here.

A first attempt to interpret the decline of internationally touring productions could lead to the assumption that because of difficult budgetary situations all over Europe right after the financial crisis, the international market to sell performing arts productions has declined too. Dramatic cuts in important neighboring countries such as The Netherlands reduced touring possibilities. That sure is a true explanation, but then how come that while internationalization decreases on the one hand, the share of international coproductions rises on the other hand? Does the crisis only affect

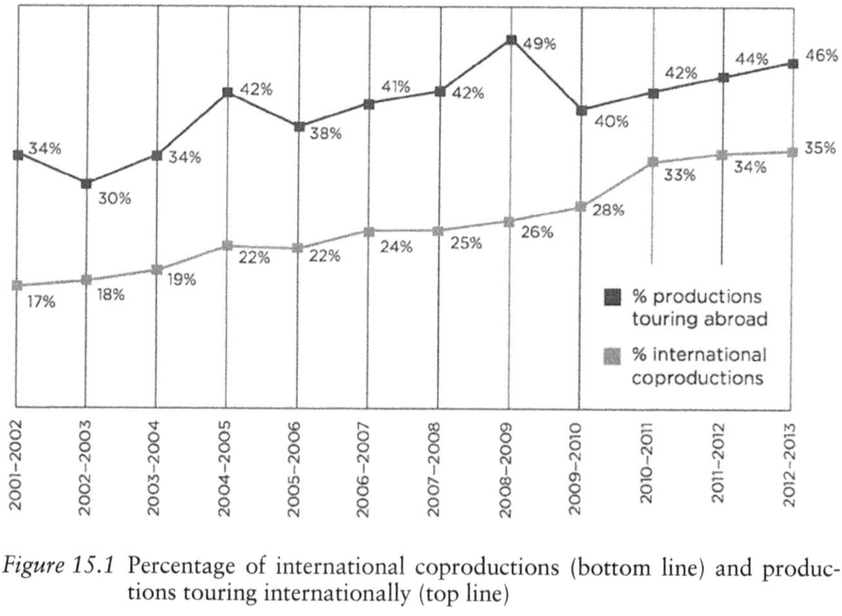

Figure 15.1 Percentage of international coproductions (bottom line) and productions touring internationally (top line)

Source: Flanders Arts Institute.

the touring side and not the production side? Is working internationally dramatically declining and booming at the same time?

A graph focusing on The Netherlands might make the case clearer (Figure 15.2). The Netherlands being traditionally an important country for Flanders in terms of international performing arts touring and collaboration, we dug deeper into this specific situation.

A very extreme case of the above mentioned combination of rising international coproductions and declining touring opportunities can be seen in this graph. The top line in 2006–2007 indicates the number of Flemish performances in The Netherlands. We see a dramatic decline of almost 30% in one year, without any signs of recovery in the following years. At the same time, the number of coproductions with Dutch organizations (the bottom line in 2006–2007) rises like never before. How can we explain these counterintuitive facts in the data? Here is where qualitative research comes in. Our proximity to the Flemish performing arts field brings us in direct contact with artists and companies. They indeed confirm that selling their productions in The Netherlands had become increasingly difficult, often as a direct consequence of budget cuts in the arts sector by the Dutch government. A way to get foreign partners aboard—and to create opportunities to disseminate artistic work—is to try to involve these organizations in the production process by setting up a coproduction.

Often, artists want to be internationally mobile because they see the benefits of the experience when making their work. They collaborate with foreign colleagues or do artistic research during a residency and they find this exchange fruitful. Although this 'positive' drive for mobility hasn't suddenly disappeared, another model has gained importance in the last years.

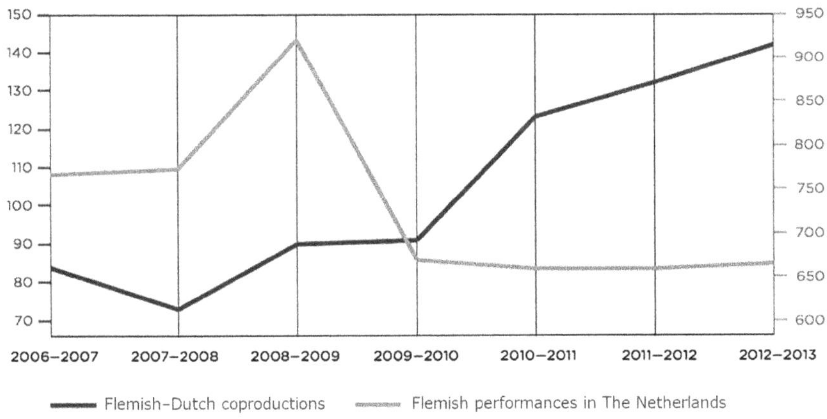

Figure 15.2 Flemish coproductions with Dutch partners (black) vs. touring opportunities in The Netherlands (grey)

Source: Flanders Arts Institute.

The economic recession brought into play a form of mobility that is not driven by artistic choice, but by economic necessity. For some artists, maximizing their international mobility is more than ever the only way to enable them to make their artistic works. It has become a *modus operandi*, a way to survive. When we take a look at the number of foreign partners—worldwide—involved in 'Flemish' productions, we see that it has quadrupled in 14 years' time. This evolution is caused both by Flemish artists seeking foreign partners and foreign artists looking for Flemish partners. For more and more productions, the combined effort and investment of more and more (also international) partners is needed to make the production possible.

The attentive reader may have noticed that when talking about 'Flemish' productions, the word 'Flemish' is always put between quotation marks. The reason is that as more productions tend to be international by nature, it gets more and more difficult to label a production in terms of a nationality. We call a production 'Flemish' whenever a Flemish partner is involved, but you could as well call it a French production when this production is coproduced with a French partner.

Just as Flemish artists and companies try to set up coproductions with foreign partners (artistically and/or economically motivated), foreign artists and companies do the same in Belgium. Going back to the example of The Netherlands, we can say that the rise of the number of coproductions between Flemish and Dutch partners is not only due to Flemish artists wanting to collaborate with Dutch colleagues. Part of the explanation is also that an increasing number of Dutch artists search for artistic collaborations in Flanders because the opportunities to do so in The Netherlands have declined. The strange, counterintuitive graph from the Netherlands now gets a lot clearer. As one of the countries that suffered heaviest from budget cuts in the arts sector, the urgent need to search for additional partners (and financial means) abroad has become very clear in the data. It explains why the number of Flemish-Dutch coproductions has never been higher, while touring opportunities in the Netherlands have fallen back dramatically.

Also, subsidized Flemish companies and artists have more and more difficulties in making their work and disseminating it internationally just on their own. They often search for additional means all over Europe and evermore worldwide. In this shifting international coproduction model, the importance of Flemish subsidies should not be underestimated. The subsidy can be seen as a sort of dowry to convince foreign coproducers, a leverage to seduce potential partners to go for a common project.

Travelogue, a European Mobility Catalogue

Another view on mobility in the performing arts leads back to 2010, when the project Travelogue was finalized (Janssens & Magnus, 2010; http://arts-mobility.info/). In this project we looked at performing arts mobility from a European perspective. Mobility had gained importance as a topic in the EU discourse from 2007 onwards. It was not less so in the EU Culture Programme. The result was a call for pilot projects to improve the lack of statistics on culture mobility at EU level. Travelogue was one of those pilot projects.

With Travelogue, we wanted to bring together the information that was available in different countries and regions. We focused on information on touring performances, as that was what was most readily available. Especially the artistic export of a country or region is typically something that authorities want to keep a record of. The information that we could assemble was interesting but very different in terms of technical formats, scope, language, and granularity. Making the information comparable thus was a challenge. Linked data technology helped us on the way. We built a flexible data model where all data on touring productions coming from different sources had to fit in. We called every statement that a certain production had toured to some place an 'observation.' Three core elements were to some extent available in all datasets and became the building blocks of each observation. The minimal requirements to talk about an observation were the presence of some artistic information, some temporal information, and some geographical information. The granularity of this information was allowed to vary. This resulted in country profiles and maps that can still be consulted on http://arts-mobility.info/.

Although the project led to very interesting conclusions in itself, the main reason to bring it up here is to mention the obstacles that we encountered during this research and how they altered our view on mobility.

First of all there is the problem of national identities of performing arts productions. In order to have clear and unambiguous data, you ideally identify a clear country of origin for every observation (e.g. a French artist making his work in France with a French company and touring abroad). When this production is 'observed' in another country—let's say Norway—you have a clear movement of a performing arts production from France to Norway. A French artistic product can be said to be exported to Norway. But an observation is always done from a certain perspective. The different datasets brought together in Travelogue lead to a myriad of perspectives. Sometimes the same 'reality' is observed differently in two datasets. With Travelogue, we managed to make observations more detailed by combining two datasets. Going back to the example of the French production touring to Norway, it might happen that a French export dataset has more detailed artistic information, whereas a Norwegian import dataset gives us more detail about the city or even exact venue where a production was staged. But very often it is not this clear-cut. Datasets might sometimes be contradictory, not necessarily because one of them contains false information, but because the perspective is different. For example: What is the country of origin of a dance production coproduced by a German and a Polish dance company with a Danish choreographer who works in Berlin? It all depends on the perspective of the observer. It turns out to be very difficult to pin down a production to one country of origin. Tracking the mobility of a production is thus not so simple as we hoped it would be.

We also realized that by looking at mobility purely from the perspective of touring productions, we missed out on a lot of interesting mobility in the performing arts. Our way of working didn't give us an insight into individual artistic exchanges or other processes that do not lead (directly) to a public performance, neither did it take into account that an artist could be mobile during workshops, residencies, and times of artistic research preceding a production shown to a public. It is important to realize that, by talking about the mobility of finished artistic products, we only get a view on that one aspect of artistic mobility. To say it in sports terminology: We didn't have any clue about the 'whereabouts' of performing artists. Although many privacy-related objections can and should be made against an evolution in that direction, it is important to know that by missing a substantial part of the picture we should not jump to conclusions too quickly.

Finally, with the data gathered in the Travelogue project, we remained somehow stuck in a quantitative approach. We managed to take some steps forward in gathering information for the EU that hadn't been available before. We showed that, with some serious efforts on the local level, a decent data collection could be set up that could deliver the EU quantitative data on the international mobility of European performing arts productions. But more than ever, we should also care about the quality of mobility. Here the Travelogue experience joins hands with the research on the international coproduction and distribution model of 'Flemish' productions above. More mobility and internationalization is not always better. It surely should never be an objective in itself. The more interesting question is what artistic mobility brings to the artists and to society at large. Frie Leysen recently pointed very beautifully to the intrinsic need for artistic mobility:

> With great solicitude, I note that art's internationalisation is coming to a halt. The dramatic cuts in The Netherlands have resulted in precisely that: the international stream is drying up. That is really dangerous! It's like a room where you shut all windows and doors. After two days, it starts to smell. It's as simple as that. You need fresh air, other colours and odours, other languages and ideas, other sensitivities.
>
> (quoted in Vidal, 2014: 52 [own translation])

This brings us to an even more central question: What is the value of arts and culture in our contemporary society and what role does cultural mobility play therein? Data on the mobility of finished art products can be one part of an answer to that question, but the data approach also falls short. It is the part that fits the dominant economic discourse of GDP, economic growth and expansion. We should be aware of not inscribing ourselves completely in that discourse, because it entails the risk for culture and the arts to become instrumentalized and not as respected in its own right as it should be. Where data falls short, it is only the artistic community that can speak up to complement that picture. Frie Leysen's quote is exemplary for what many artists feel in this respect. As a culture sector, it is more than ever our duty to inject the discourse with qualitative arguments about the value of artistic mobility for contemporary society.

Or, as it was stated in our 2014 field analysis:

> The public function of art is fundamentally uninstrumental, but originally operates by its own logic and according to its proper rules. Picasso got famous because he looked at things differently, not because he wanted to stimulate tourism in Paris. Similarly, Anne-Teresa De Keersmaeker conquered the world because she saw dance in a new perspective, not because she wanted to become a Flemish export product.
>
> (Kunstensteunpunt vzw, 2014: 5 [own translation])

Again, Einstein was right. Yes, look at the numbers. Look at them carefully, but do not let them blind you.

References

Albert Einstein quote on *The Quotations Page*. Available at: www.quotationspage.com/quote/26950.html Accessed 20 June 2016.

Cavyn, F., De Moor, M., Hesters, D., Janssens, J., Magnus, B., and Wellens, N. (2011). *Ins & Outs: A Field Analysis of The Performing Arts in Flanders*. Available at: http://vti.be/nl/over-vti/publicaties/ins-outs-field-analysis-performing-arts-flanders Accessed 20 June 2016.

Janssens, J. (Ed.) (2014). *Transformers: Landscape Sketch for the Performing Arts from Flanders and Beyond*. Available at: http://vti.be/nl/over-vti/publicaties/publication-transformers-landscape-sketch-performing-arts-flanders-and-beyond Accessed 20 June 2016.

Janssens, J. (2016). *After the Growth? Trends in the International Production and Distribution of the Performing Arts*. Available at: https://issuu.com/kunstenpuntflandersartsinstitute/docs/flanders_arts_institute_after_the_g Accessed 20 June 2016.

Janssens, J., and Magnus, B. (2010). *Travelogue: Mapping Performing Arts Mobility in Europe*. Available at: http://arts-mobility.info/ Accessed 20 June 2016.

Janssens, J., and Moreels, D. (2007). *Metamorphoses: Performing Arts in Flanders and Brussels Since 1993*. Available at: http://vti.be/nl/over-vti/publicaties/metamorphoses-performing-arts-flanders-1993 Accessed 20 June 2016.

Kunstensteunpunt vzw. (2014). *Landschapstekening Kunsten*. Available at: http://vti.be/sites/default/files/Landschapstekening%20kunsten.pdf Accessed 20 June 2016.

Lilley, A., and Moore, P. (2013). *Counting What Counts: What Big Data Can Do for the Cultural Sector*. Available at: www.nesta.org.uk/publications/counting-what-counts-what-big-data-can-do-cultural-sector Accessed 20 June 2016.

Vidal, K. (2014). Kunstenaars zijn zo braaf en volgzaam. *De Morgen*, 3 May 2014, 52.

Index

Note: Page numbers in *italics* indicate figures.

For Product Safety Concerns and Information please contact our EU representative GPSR@taylorandfrancis.com
Taylor & Francis Verlag GmbH, Kaufingerstraße 24, 80331 München, Germany

www.ingramcontent.com/pod-product-compliance
Ingram Content Group UK Ltd.
Pitfield, Milton Keynes, MK11 3LW, UK
UKHW011013180926
475701UK00004B/713